T0360493

ROUTLEDGE LIBRARY EDITIONS: URBANIZATION

Volume 8

URBANISATION AND LABOUR MARKETS IN DEVELOPING COUNTRIES

URBANISATION AND LABOUR MARKETS IN DEVELOPING COUNTRIES

STUART W. SINCLAIR

Routledge
Taylor & Francis Group

LONDON AND NEW YORK

First published in 1978 by Croom Helm Ltd

This edition first published in 2018
by Routledge
2 Park Square, Milton Park, Abingdon, Oxon OX14 4RN

and by Routledge
605 Third Avenue, New York, NY 10017

Routledge is an imprint of the Taylor & Francis Group, an informa business

© 1978 Stuart W. Sinclair

All rights reserved. No part of this book may be reprinted or reproduced or utilised in any form or by any electronic, mechanical, or other means, now known or hereafter invented, including photocopying and recording, or in any information storage or retrieval system, without permission in writing from the publishers.

Trademark notice: Product or corporate names may be trademarks or registered trademarks, and are used only for identification and explanation without intent to infringe.

British Library Cataloguing in Publication Data
A catalogue record for this book is available from the British Library

ISBN: 978-0-8153-8014-6 (Set)
ISBN: 978-1-351-21390-5 (Set) (ebk)
ISBN: 978-0-8153-7841-9 (Volume 8) (hbk)
ISBN: 978-1-351-23243-2 (Volume 8) (ebk)

Publisher's Note
The publisher has gone to great lengths to ensure the quality of this reprint but points out that some imperfections in the original copies may be apparent.

Disclaimer
The publisher has made every effort to trace copyright holders and would welcome correspondence from those they have been unable to trace.

URBANISATION AND LABOUR MARKETS IN DEVELOPING COUNTRIES

SOME OTHER ELBS LOW-PRICED EDITIONS

Beckerman	AN INTRODUCTION TO NATIONAL INCOME ANALYSIS	*Weidenfeld & Nicolson*
Bridger and de Soisson	FAMINE IN RETREAT?	*Dent*
Cox, Healey and Moore	BIOGEOGRAPHY	*Blackwell Scientific*
Epstein	SOUTH INDIA: YESTERDAY, TODAY AND TOMORROW	*Macmillan*
Hicks	VALUE AND CAPITAL	*Oxford University Press*
Hunter, Bunting and Bottrall	POLICY AND PRACTICE IN RURAL DEVELOPMENT	*Croom Helm*
Meade	THEORY OF INTERNATIONAL ECONOMIC POLICY, Vols. I & II	*Oxford University Press*
Morton and Tulloch	TRADE AND DEVELOPING COUNTRIES	*Croom Helm*
Moser and Kalton	SURVEY METHODS IN SOCIAL INVESTIGATION	*Heinemann Educational*
Prest	PUBLIC FINANCE IN DEVELOPING COUNTRIES	*Weidenfeld & Nicolson*
Robinson	BIOGEOGRAPHY	*Macdonald & Evans*
Silverman	THE THEORY OF ORGANISATIONS	*Heinemann Educational*
Thirlwall	GROWTH AND DEVELOPMENT	*Macmillan*

URBANISATION AND LABOUR MARKETS IN DEVELOPING COUNTRIES

STUART W. SINCLAIR

THE ENGLISH LANGUAGE BOOK SOCIETY
AND
CROOM HELM   LONDON

© 1978 Stuart W. Sinclair
Croom Helm Ltd, 2-10 St John's Road, London SW11

ELBS edition first published 1979

ISBN 0-7099-0075-9

To Winifred Doreen Sinclair
and Eric Fraser Sinclair

Printed in Great Britain by
Biddles Ltd, Guildford, Surrey

CONTENTS

LIST OF TABLES

PREFACE

There are a number of anthropological, sociological and geographical studies of squatter settlements and urbanisation in underdeveloped countries, but relatively few by economists. Now that 'development studies' has gained a degree of academic respectability, reasserting the divisions between disciplines seems a little pedantic, but there are differences in approach and emphasis to be discerned. While the anthropologist may, for instance, find the concepts of 'slums of hope' and 'slums of despair' illuminating in explaining shanty dwellers' behaviour and aspirations, a more economic perspective will be less concerned with urban dwellers' perceptions of their condition than with their behaviour as they arrive in cities, move about inside them, look for jobs or make plans to return home. Clearly it would be tedious to attempt a list of 'economic' considerations distinct from other types; it would also be wrong to imply that one approach is in any sense 'better' than another. What is being assumed is rather an approach to shanty towns and squatter settlements which is different from those of other disciplines and which can hope to present old wine in new bottles at the very least, and perhaps find some new vintages as well.

What economists have so far contributed to the study of labour in cities in underdeveloped countries (UDCs) is often scattered in inaccessible journals or unpublished reports. By drawing together as much of this material as possible, it is hoped that this book will allow the undergraduate in geography, economics or development studies to grasp a complex literature more easily. The appeal to a varied readership is not, hopefully, at the expense of trying to be all things to all men. Rather, an attempt has been made to include some material familiar to all development economists and some which they are unlikely to know. If the treatment of the issues presented here is cause for dissatisfaction, this is welcome, for more work badly needs to be done. As a recent review of literature on urbanisation concluded, there is 'very little reference to the work situation and its place in the urbanisation process . . . nowhere is it as central to the analysis . . . as one might expect'.[1] This book attempts to fill that gap, by setting out the major controversies relating to urban labour markets in UDCs. It pays particular attention to work which goes on outside large-scale firms. For that reason labour absorption and earnings within large firms are dealt with briefly.

Among the many people who have helpfully commented on drafts of this book are Kaushik Basu, Richard Disney, Alistair Fischer, Lindsey Hughes, Matthew McQueen, Guy Routh and Alan Udall. All remaining inadequacies of analysis and infelicities of expression are, of course, my own.

Notes

1.　J. Vincent, 'Urbanization in Africa: A Review Article', *Journal of Commonwealth and Comparative Politics*, Vol. 14, No. 3, November 1976, pp. 286-98. The quotation is from p. 292.

1 THE GROWTH OF CITIES: A DESCRIPTION

The subject matter of this book is necessarily complicated: as complicated as the life of a city. To make comprehensible the mass of detail, the diversity of experience and the intricacy of relationships, it is obviously necessary to simplify and to generalise, and in economics simplification by theorising is accepted as both useful and necessary. To deal with 837 cities in UDCs with a population of over 100,000 it is essential to generalise if analysis is not to be conducted wholly on an anecdotal level. But such systematising of evidence will inevitably do some damage to the data; and certain further difficulties arise in development studies. Different countries have widely differing practices, ideologies, even climatic conditions. A pattern of landholding espoused in one country may be totally inappropriate elsewhere — because of labour supplies, technology options, the nature of the crops, or the risk preference of the workers. And what represents a solution for an unemployment problem in one country might be at best a palliative and at worst might exacerbate the problem in another setting. Clearly, universality of approach — the appeal of the Grand Theory — must be tempered. Social scientists have often tended to devise theories and analytical tools with pretensions to universal relevance, believing this to be the practice of natural scientists. They are thus to be twice damned — once for ignorance and once for envy.[1] Generalisations about cities, and therefore countries, will appear in the following pages. What is desired is an overview of the patterns of labour-use emerging in UDCs' cities; and some local nuances will have to be sacrificed to this end.

This chapter begins with some data on the extent and nature of world urbanisation. These figures provide a preliminary guide to the world's recent population movements, which in turn are the substance of the migration patterns discussed in chapter 2. After these tables are presented and discussed, some conceptual issues are raised. It is argued that a satisfactory treatment of the economics of labour in cities and shanty towns demands the rejection or modification of a number of the terms conventionally used in development studies.

Table 1.1 reproduces data on global urbanisation, which, naturally, can be only an approximate guide. The difficulties of estimating cities' sizes are both logistical and definitional. In UDCs, the recent growth

13

Table 1.1: World Urbanisation,[a] 1800—1970

Year	Urban %	% in cities of over 100,000
1800	3.0	1.7
1850	6.4	2.3
1900	13.6	5.5
1950	28.2	16.2
1970[b]	38.6	23.8

[a] The criterion is places with a population in excess of 5,000.
[b] Projected from 1968.

Source: K. Davis, *World Urbanization 1950—1970*, Vol. 1 (Berkeley: University of California Press), p. 10.

of towns and cities in unplanned and haphazard shanty towns will make counting extremely hard. One might proceed by counting the number of huts or houses and multiplying by some factor (five, ten or twenty, say), to estimate total population; alternatively, one might count as accurately as possible the population of a small area, then multiply this by the estimated area of the whole town. The tendency for the rates of migration, including the trend of reverse migration out of towns, to change seasonally or irregularly, naturally exacerbates the problems of estimating. The last census of Nigeria was declared null by the new 1975 government, since it believed that the problems of accurate quantification had allowed some groups in the country to exaggerate their size. The absence of consistent birth and death records is a further handicap. Despite these obstacles, Table 1.1 shows some rough trends. The growth of urbanisation during and after the industrial revolutions of the presently rich countries, and in particular the growth of exceptionally large cities, is evident.

Table 1.2 provides data for urban growth in UDCs, indicating growth rates nearing 10 per cent per year in 1950-75, but which, in most cases, will subside over 1975—2000. The last column gives an impression of the vastness of some UDCs' cities as anticipated by the end of the century. But this same vastness highlights the second difficulty in estimating urbanisation — that of defining a city. Naturally a city of 31.5 million people — as Mexico City is predicted to be — will be more in the nature of a system of cities and towns.

Table 1.2: Populations of Selected Urban Areas, 1950–2000
in millions

Country	1950	Average annual rate of growth	1975	Average annual rate of growth	2000 (estimated)
Underdeveloped Countries		%		%	
Mexico City	2.9	5.4	10.9	4.4	31.5
Buenos Aires	4.5	2.9	9.3	1.5	13.7
Sao Paulo	2.5	5.7	9.9	3.9	26.0
Rio de Janeiro	2.9	4.4	8.3	3.4	19.3
Bogota	0.7	6.5	3.4	4.2	9.5
Cairo	2.4	4.3	6.9	3.6	16.9
Seoul	1.0	8.3	7.3	3.8	18.7
Manila	1.5	4.4	4.4	4.3	12.8
Kinshasa	0.2	9.7	2.0	5.6	7.8
Lagos	0.3	8.1	2.1[a]	6.2	9.4
Shanghai	5.8	2.8	11.5	2.6	22.1
Peking	2.2	5.8	8.9	3.7	22.0
Jakarta	1.6	5.1	5.6	4.7	17.8
Calcutta	4.5	2.4	8.1	3.7	20.4
Bombay	2.9	3.7	7.1	4.2	19.8
Karachi	1.0	6.2	4.5	5.4	16.6
Developed Countries					
New York	12.3	1.3	17.0	1.3	22.2
London	10.2	0.2	10.7	0.7	12.7
Paris	5.4	2.1	9.2	1.2	12.4
Tokyo	6.7	3.9	17.5	2.0	28.7

[a] The latest estimate for Greater Lagos is 3.3 million.

Source: IBRD, 'The Task Ahead for the Cities of the Developing World'
(Washington DC: World Bank Staff Working Paper No. 209, 1975), p. 20.

Just like New York, London and the major capitals of rich countries
now, there will be many more or less parochial sub-centres, self-
supporting in some respects, interrelated in others. Related to this is
the point that as cities grow to touch previously outlying settlements,
the settlements' entirety is included in the expanding city's growth

Table 1.3: Slums and Uncontrolled Settlements: Percentage of Total
Population in Cities

1970 GNP/ capita ($ US)	City	Country	%	Year
980	Caracas	Venezuela	40	1960-6
920	Singapore		15	—
730	Panama City	Panama	17	—
720	Santiago	Chile	25	—
670	Mexico City	Mexico	46	—
670	Kingston	Jamaica	25	—
590	Beirut	Lebanon	15	—
450	Lima	Peru	40	—
420	Rio de Janeiro	Brazil	30	—
400	Lusaka	Zambia	58	1964
380	Kuala Lumpur	Malaysia	37	1947-57
360	Guatemala City	Guatemala	30	—
340	Bogota	Colombia	60	—
320	Baghdad	Iraq	29	—
310	Istanbul	Turkey	40	1960-5
310	Abidjan	Ivory Coast	60	1955-63
310	Accra	Ghana	53	—
290	Guayaquil	Ecuador	49	—
280	Tegucigalpa	Honduras	25	—
250	Amman	Jordan	14	—
250	Seoul	Korea	30	1955-65
240	Monrovia	Liberia	50	—
230	Casablanca	Morocco	70	—
230	Dakar	Senegal	60	—
210	Manila	Philippines	35	—
180	Douala	Cameroon	80	—
150	Nairobi	Kenya	33	1963-73
140	Lomé	Togo	75	—
110	Calcutta	India	33	—
110	Colombo	Sri Lanka	43	—
100	Karachi	Pakistan	23	—
100	Dar es Salaam	Tanzania	50	—
90	Kinshasa	Zaire	60	—
80	Kabul	Afghanistan	21	—
80	Addis Ababa	Ethiopia	90	—
80	Blantyre	Malawi	56	—
80	Katmandu	Nepal	22	—
80	Djakarta	Indonesia	26	—
70	Mogadishu	Somalia	77	—
60	Ouaga-dougou	Upper Volta	70	—

— = not available.
All estimates are extremely rough.

Source: O.F. Grimes, Jr., *Housing for Low-Income Urban Families* (London:
Johns Hopkins Press for IBRD, 1976), Table A2, pp. 118-27.

statistics in one fell swoop, presenting an illusory rate of increase. Note in this table the slower rates of growth predicted for cities in the developed countries. More recent analyses expect absolute declines in the size of some cities: in the USA, the population of the Standard Metropolitan Areas is expected to continue its post-1970 decline of approximately 0.4 per cent annually.[2] Table 1.3 shows how far the recent growth of UDCs' cities has resulted in uncontrolled settlements and shanty towns. Again, the proviso of judicious interpretation of these statistics is necessary. Some cities, it can be seen, have provided, either through public or private housing, for most of their expanding populations: Singapore, for instance, has only 15 per cent of its population classified as uncontrolled. The large percentages of slum housing may overstate the development problems in some UDCs, however, since many inhabitants of so-called uncontrolled settlements have steady jobs and possess a number of consumer durables – TVs, radios, and so on – which indicate that they do not live in the severest deprivation. This comment will be elaborated upon in chapters 4, 5 and 6, where the assumed equivalence of 'uncontrolled' housing and labour markets with poverty is examined.

Table 1.4 shows how far migration has contributed to the growth of population in selected UDCs' cities. Only in Taipei (43 per cent) and Bogota (33 per cent) are the proportions less than 50 per cent. The population increase of existing residents accounts for the rest.

The factors traced in these tables conspire to make urban life treacherous and mean for many migrants. The case of one city – Lagos, Nigeria – is examined now to illuminate the results of the urbanisation just outlined

The total population of metropolitan and fringe Lagos is unknown, but was thought to be 1.2 to 1.5 million in 1972,[3] and about 3.3 million in 1976.[4] Its growth has been estimated at 19 per cent per year.[5] But naturally these figures must be interpreted with great care. Chief among the difficulties of assessing the population of a town like Lagos is the presence of non-Nigerians who drift to the capital seasonally, or in response to a disaster such as the Sahelian drought. It appears that urbanisation conspires to lower the reproduction rate: 'Better-educated, urban women tend to be young, marry later, begin their child-bearing at a later age, and plan to have fewer children.'[6] (The average age of marriage in Western Nigeria and Lagos, at 19.4, is appreciably higher than that for the rural North, at 14.8.[7]) Fifty per cent of the population of Lagos is under 15 years of age,[8]

Table 1.4: Migrants as a Percentage of Recent Urban Population
Increases

City	Period	Total population increase (thousands)	Migrants as a percentage of total population increase
Abidjan	1955-63	129	76
Bogota	1956-66	930	33
Bombay	1951-61	1,207	52
Caracas	1950-60	587	54
	1960-6	501	50
Djakarta	1961-8	1,528	59
Istanbul	1950-60	672	68
	1960-5	428	65
Lagos	1952-62	393	75
Nairobi	1961-9	162	50
Sao Paulo	1950-60	2,163	72
	1960-7	2,543	68
Seoul	1955-65	1,697	63
Taipei	1950-60	396	40
	1960-7	326	43

Source: IBRD, *Urbanization* (Washington DC: World Bank Sector Working Paper, 1972), p. 80.

and this has a major effect on overcrowding of houses. In 1966, the median household contained 5.5 persons, but in 1972 it had risen to 7.8.[9] From Table 1.5, which shows the distribution of household density between the districts of Lagos, it is noticeable that the city-centre, especially Lagos Island, is the most crowded, and also experienced the greatest increase in population density over the period 1962-72. The more outlying parts of the city, such as Yaba and Surulere, became more crowded too, but less markedly. The case of Ikeja — 13 miles from Lagos Island — is altered by the siting of a new industrial estate and Murtala Mohammed Airport: these tend to attract more people than would otherwise have been expected. Living conditions in the face of such overcrowding have, not surprisingly, deteriorated. A Ford Foundation survey commented:

. . . chaotic traffic conditions have become endemic; demands
on the water supply system have begin to outstrip its maximum

Table 1.5: Household Population Density in Selected Lagos Districts, 1966 and 1972 (persons per household)[a]

District	1966	1972	Change (%) 1966-72	Proportion of non-Nigerians*
Victoria Island	—	4.8	—	50 %
Ikoyi	9.0	6.1	−32.2	
Obalende	5.5	8.7	58.1	
Lagos Island	5.4	10.8	100	13 %
Ebute-Metta	5.03	7.4	46	
Yaba		7.3		
Apapa	7.0	5.6	−20	
Ajegunle		5.8		
Surulere	5.8	7.8	34.4	
Mushin/Idioro	5.0	8.0	60	
Ikeja	3.8	5.4	42.1	72 %
Palmgrove	—	—	—	47 %
ALL:	5.4	7.8	44.4	

[a] A more recent report estimates household density at 32 persons per house on Lagos Island. See Economist Intelligence Unit, *Quarterly Economic Review of Nigeria* (London: The Economist, 1977), No. 2, p. 10.

Sources: T.M. Yesufu, 'Characteristics and Changes of the Lagos Population', Table 4, p. 6; * T.M. Yesufu, 'Population, Employment and Living Conditions in Lagos', pp. 8-9.

capacity; power cuts have become chronic as industrial and domestic requirements have both escalated; factories have been compelled to bore their own wells, and to set up stand-by electrical plants; public transport has been inundated; port facilities have been stretched to their limits; the congestion of housing and land-uses has visibly worsened, and living conditions have degenerated over extensive areas within and beyond the city's limits, in spite of slum clearance schemes; and city government has threatened to seize up amidst charges of corruption, mis-management and financial incompetence.[10]

A journalist's impressions are still more vivid:

Not so long ago, the United Nations called Lagos the dirtiest capital in the world. Since then, only feeble attempts have been made to clean up Africa's fastest-growing city. It now chokes on bad housing, traffic jams and cement pile-ups. Whole streets are littered with compost, broken bottles, pieces of furniture, empty cans. The stench is intolerable during the dry season and worse when it rains. Then, over half the streets are flooded. But the city-dwellers have learnt to live with it all. Street-parties — owambes — are still held in the evenings. The children don't escape though. About a seventh of all babies die before their first birthday. Infant mortality is usually due to respiratory and infectious diseases. Food stalls are sometimes sited near rubbish-dumps . . .[11]

The housing conditions in this milieu are scarcely a refuge. Twenty-two per cent of Lagos households have no access at all to pipe-borne water, and a further 15 per cent have it in the neighbourhood.[12] Sixty per cent of houses have no modern lavatory, and the 40 per cent with flush toilets are concentrated in the expatriate and high-income areas of town. Eighty-two per cent of houses along Lagos Lagoon have no public refuse disposal. Among the results is the fact that 85 per cent of Lagos schoolchildrem have either hook-worm or round-worm; and that 10 per cent of deaths are attributable to dysentery or diarrhoea.[13] Attempts to improve the housing stock have been leisurely: house-building proceeded at only 5.1 per cent annually in the late 1960s. The 15,000 housing units built for the festival of Black Arts and Culture, to which Lagos was the host in early 1977, will alleviate conditions a little, but only if they are rented at reasonable rates. In the private rental sector, large houses rent for ₦ 20,000 a year and more (₦ 1, one naira, being equivalent to $ 1.00 or £0.70), and renters frequently have to pay three years' rent in advance. At the low-income range, a 1972 survey established that 'in spite of the low standard of housing, and government intervention in the form of rent control edicts, 70 per cent of the households in the low-income areas pay over ₦ 10.00 per month in rent, and 30 per cent pay more than ₦ 20.00. This represents a substantial increase over the 1968 figure'.[14] (In 1975 an unskilled government worker would expect to receive a wage of ₦ 720.00 annually.) Due to the high cost of buying and people's frequent preference for renting houses in cities where they work and buying instead in their home town or village, 79 per cent of people in Lagos rent their accommodation: 'The majority of migrants want cheap rented rooms rather than a house in town; any money they save is spent on a house at home.'[15]

Within the city, youngsters arrive from all over the country —
exceptionally, even from neighbouring countries — to find a job, to
go to school, or to be with friends and relatives. Their chance of finding
a job will depend on how much education they have, how many
friends they have — and where — and on luck. In the belief that it
provides security, a guaranteed income, and perhaps a prestige not
conferred by other jobs, a large number of these youngsters attempt
to enter local or federal government employment. If they fail, as
most of them must, they will try the larger factories and workshops,
eventually trying their luck in the 'urban lottery' that is the so-called
trade-service sector. Plying queueing cars with cassettes, tissues, soap,
magazines and tins of food along the choked bridges of Lagos Island,
or running errands for a shopkeeper, or learning a trade by apprenticeship — their income is mercurial. Assistance from family or friends
may be forthcoming for a while, but that will depend on the family's
wealth, disposition to visitors, or their estimation of the charge's
future earnings. All in all, the young arrival in Lagos faces a dismal
trudge in life and work.

The Economics of Urbanisation

The preceding overview of the effects of 'urban imbalance'[16] already
suggests that conventional economic theory may have to be modified
to cope with the experience of cities in UDCs. Development
economics has already discarded many of the more vigorous assumptions of classical economic theory: those of perfect factor mobility
and of perfect knowledge; those relating to human capital theory,
where there is no market for borrowing educational funds; and of
homogeneity of labour. But some other modifications must be made
if a useful vocabulary of urban development is to be constructed.
Primarily, the modifications affect concepts of employment, wages
and housing.

John Weeks was among the first to question the relevance of
employment as a criterion of development, or even as a criterion of
labour-use.[17] Believing most economists' use of the term to be
ethnocentric, Weeks argued that other cultures' patterns of work and
labour-utilisation were being forced into the inappropriate conceptual
framework of Western unemployment and employment. Quantifying
unemployment had, he argued, become the uncritical procedure of
international agencies, notably the International Labour Organisation
(ILO). Related to this was the long debate on the productivity of
labour in the unregulated agricultural sector of UDCs. Since the

evolution and elaboration of the celebrated Lewis model,[18] a theoretical and empirical debate has continued over whether the marginal productivity of labour in agriculture and elsewhere was zero (that is, whether or not labourers could be removed while leaving output unchanged). Those who answered that this was the case used the terms 'underemployment' and 'hidden unemployment' to describe this phenomenon, but again it was felt that these terms were unsatisfactory:

. . . the shift from this sterile and involuted debate to the discussion of the 'informal sector' was a welcome relief and a considerable advance. This new formulation entailed a renunciation of the morbid fascination with the measurement of the exact size of the unutilized labour pool, and focused attention on the real issue — which was the role played by the small-scale, less-regulated producers.[19]

The simple replacement for employment and unemployment was the level of earnings. This introduces its own problems, of course. A portion of the subsistence of the urban poor will never pass through a market: it will consist of food or lodgings provided for services rendered; their own services provided in lieu of rent; or the obligations of the family network to support newcomers while job-seeking or going to school. Income too is unlikely to be recorded by many of the individuals or enterprises on the fringes of the cash economy; nor will one be able to gauge income as a rate, measured by days or months. But, whatever its shortcomings, the use of 'income' to embrace notional as well as cash subsistence is an improvement over 'employment', for it allows recognition of the irregular nature of the work, be it for payment in kind, or undertaken in haphazard hours, in which many of the urban poor are likely to be involved. Throughout this book, the term 'employment' will be used only to refer to work in the 'formal' sector, where wages are paid in money for regular hours performed; in other occupations, the less rigorous terms 'earnings' and 'subsistence' will be used.

Next, this book will not make use of a distinction sometimes made by urban planners and sociologists between slums and squatter settlements. There is, after all, no reason why this distinction should be maintained if it is not appropriate; it is merely a heuristic device which should not be allowed to live a life of its own. The distinction is usually phrased thus: the slum settlement is legal and pays rents. The squatter settlement is illegal, its inhabitants are predominantly new arrivals from the countryside, and the shelter they build is felt both by them and by the authorities to be merely temporary.[20] Useful characteristics have been ascribed to both the squatter settlement and the slum: 'These . . . are rightly termed "improving" spontaneous settlements, or settlements of hope; they are not dens of crime, vice and disease, but contain enterprising, well-organized, responsible citizens, who by community self-help have accomplished striking improvements in their condition.'[21]

Stokes' seminal essay, on the other hand, contrasts 'slums of hope' and 'slums of despair'.[22] This distinction stresses the residents' different perceptions of their chances of social and economic mobility. Residents of 'slums of hope' believe their poverty will only be a transient phase in their urban experience, which would be surpassed as their savings allowed. 'Slums of despair', however, are thought to be inhabited by those who feel their chances of improvement to be very slight and outside their immediate control. They tend to respond to opportunity with some doubt, and not to recognise opportunities as readily as those in 'slums of hope'. This is an illuminating distinction for some purposes, as indeed is Turner's parallel distinction between 'bridgeheaders' (who live close to their work and in insecure circumstances) and 'consolidators' (who live further from the city-centre, and invest in buying and furbishing a house).[23] But these typologies cannot offer much help for labour market analysis in UDCs' cities. The aspirations and perceptions of urban workers unquestionably play an important role in determining their behaviour at work, as well as the way they seek work, but this is more the subject matter of industrial relations. This study of labour markets will skirt these complications, and study the expression of preferences rather than their source. Drakakis-Smith's survey of Asian urban renewal set out with a four-part typology of low-cost housing, as reproduced in Figure 1.1. But he found it convenient to abandon the distinction he devised between squatter and informal slum housing, primarily because many people who chose to live in slums were relatively affluent.[24] His conclusion, that 'more consideration has to be given to the microeconomics of the informal sector and the need to retain as many of the residential-economic ties as possible', confirms that the slum/squatter settlement distinction should not be adopted hereafter.

The procedure in the following chapters, then, is to approach the concepts of wages, employment and jobs with circumspection, and to sidestep the distinction drawn by others between slums and squatter settlements. Chapter 2 now goes on to examine migration and

Figure 1.1: Four-Part Typology of Low-Cost Housing

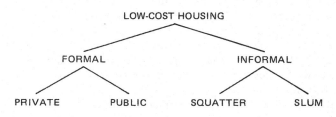

population growth in some detail. The varieties of migration found in UDCs are discussed, as are their various impacts on the economic life of cities. This leads to the discussion, in chapter 3, of initial behaviour in UDCs' cities. The task of job-seeking is introduced, as are the roles of such institutions as trade unions and trade associations. The failure of manufacturing employment in large-scale firms to absorb this burgeoning labour force is studied in chapter 4, as are the alternative livelihoods to which people have turned in consequence. The so-called 'informal' sector, in its various guises, is examined in chapters 5 and 6; and the final chapter draws together the threads scattered in the main chapters and reviews the book's main conclusions.

This introductory chapter has shown the importance of slums and shanty towns in terms of world population. It has also discussed the type of economic analysis suited to the study of these towns. Certain parts of conventional theory have to be modified; it is necessarily an eclectic approach. The following chapter turns to the growth of towns in UDCs, examining more closely how migration has come about since 1945. This sets the scene for an examination, in subsequent chapters, of labour market activities within towns and cities.

Notes

1.   This point is neatly made in J.M. Culbertson, *Macroeconomic Theory and Stabilization Policy* (New York: Wiley, 1971), p. 110.
2.   B.J.L. Berry (ed.), *Urbanization and Counter-Urbanization* (Beverly Hills: Sage Publications, 1976), p. 24.
3.   T.M. Yesufu, 'Characteristics and Changes of the Lagos Population' (University of Lagos Human Resources Research Unit, Bulletin 2/001, 1972), p. 2.
4.   *African Development* (London), December 1976, p. 1319.
5.   Ford Foundation, *Urbanization in Nigeria* (New York: Ford Foundation

International Urbanization Survey, 1972), p. 6.
6. F.L. Mott and O.J. Fapohunda, 'The Population of Nigeria' (University of Lagos Human Resources Research Unit, Research Monograph No. 3, 1975), p. 37.
7. Ibid., p. 35.
8. Yesufu, p. 2.
9. T.M. Yesufu, 'Population, Employment and Living Conditions in Lagos' (University of Lagos Human Resources Research Unit, Bulletin 2/002, 1972), p. 6.
10. Ford Foundation, *Urbanization in Tropical Africa* (New York: Ford Foundation International Urbanization Survey, 1972), p. 37.
11. *African Development*, December 1976, p. 1319.
12. Yesufu, Bulletin 2/002, p. 11.
13. Ibid., p. 14.
14. Ibid., p. 25.
15. M. Peil, 'African Squatter Settlements: A Comparative Study', *Urban Studies*, Vol. 13, No. 2, June 1976, pp. 155-66.
16. The term is John Weeks'. See 'An Exploration into the Nature of Urban Imbalance in Africa', *Manpower and Unemployment Research in Africa*, Vol. 6, No. 2, November 1973, pp. 9-36.
17. J. Weeks, 'Does Employment Matter?', *Manpower and Unemployment Research in Africa*, Vol. 4, No. 1, 1971, pp. 4-10.
18. W.A. Lewis, 'Economic Development with Unlimited Supplies of Labour', *Manchester School*, Vol. 22, No. 2, 1954, pp. 139-91.
19. M.A. Bienefeld and E.M. Godfrey, 'Measuring Unemployment and the Informal Sector', *Institute of Development Studies Bulletin*, Vol. 7, No. 3, October 1975, p. 7.
20. M. Juppenlatz, *Cities in Transformation: The Urban Squatter Problem of the Developing World* (Queensland: University of Queensland Press, 1970), pp. 5, 12-13.
21. A.B. Mountjoy, 'Urbanization, the Squatter and Development in the Third World', *Tijdschrift voor economische en sociale geografie*, Vol. 67, No. 3, 1976, pp. 130-7.
22. C.J. Stokes, 'A Theory of Slums', *Land Economics*, Vol. 38, 1962, pp. 187-97.
23. J.F.C. Turner, 'Housing Priorities, Settlement Patterns and Urban Development in Modernizing Countries', *Journal of the American Institute of Planners*, Vol. 34, 1968, pp. 354-63.
24. D.W. Drakakis-Smith, 'Urban Renewal in an Asian Context: A Case-study of Hong Kong', *Urban Studies*, Vol. 13, 1976, pp. 295-305.

2 POPULATION AND MIGRATION TO CITIES

'While the causes of the worldwide fall in death rates are well understood, the interrelations among the socioeconomic processes related to fertility — education, health, urbanisation, mobility, etc. — are still subjects of active debate.'[1] This summary, which accompanied a 1970 World Bank research report on similarities in international patterns of development, would be accepted by most demographers. There would be some debate over the role of economic development up till now in UDCs in reducing birth rates, but most would agree that medical science developed in the developed countries (DCs) is a prime explanation. 'A Stone Age people can be endowed with a low twentieth-century death rate within a few years, without waiting for the slow process of economic development or social change.'[2]
In Latin America,

> Mortality rates fell dramatically as a result of the introduction of modern health techniques . . . after 1915, principally by the Rockefeller Foundation and the Pan American Health Organization . . . The result was that in a country like Costa Rica, mortality rates fell by 50% within 20 years, — a reduction which took Britain, starting in the 18th Century, over 150 years to achieve. In Costa Rica . . . the death rate is 8 per 1,000 population, lower than the 9.6 per 1,000 of the USA. On the other hand, the birth rate in Costa Rica is over two and a half times that of the USA . . . Indeed, according to some calculations, Costa Rica's population growth rate could be set as high as 4.5% a year, while the death rate per 1,000 population may be the even lower figure of 7. Within a century, its present population of around 1.8 million, given the current rate of advance, may increase to 75 million, 44 times the present total.[3]

This population upsurge — a low death rate combined with a high birth rate — is occurring at a much lower level of income per capita than was the case in developed countries. 'In short, today's non-industrial populations are growing faster, and at an earlier stage than was the case in the demographic cycle that accompanied industrialization in the nineteenth century.'[4]

That a relationship between birth rate and per capita GNP exists is beyond doubt; but this simple statistical relationship explains little, of course. One has to introduce more complex variables or proxies in order to establish the impact of such related phenomena as educational attainment (the extent and spread of education), and urbanisation. Higher education certainly induces lower birth rates in UDCs, and urbanisation sometimes does. The effects of these two are hard to distinguish in practice because enrolment ratios, expenditure per pupil, and other measures of 'education' are themselves correlated to the degree of a country's urbanisation. While the more educated in UDCs certainly exhibit falling birth rates, there is a substantial lag before this is transmitted to the less educated. Children are still perceived as ultimate providers of security for the parents.[5]

Data from South-East Asia suggests that while the poor may delay marriage and childbearing simply because they cannot afford it, it is largely among the poorest that the fastest rates of increase are found.[6] Elsewhere, the patterns of population increase depend on cultural factors, too. Chapter 1 noted different average ages of marriage between the rural Hausa and the more urban Yoruba of Nigeria; there are also factors like the shorter nursing period exhibited by women in towns, which shortens the time during which there are taboos against intercourse.[7] Furthermore, the precise causal link between education and number of children desired is unclear. One interpretation is that the opportunity cost of bearing children is increased, since the more educated person's earnings foregone will be greater than those of a less educated person.[8] Rising population in UDCs increases the pressure on cultivable land. If the area of land under cultivation is not increased, and the productivity of that land is constant, too, food consumption per capita will fall unless some people cease to be dependent upon the produce of the area. The implications of rising population for migration are therefore related to the trends in rural output: if output is falling or stagnant there will, other things being equal, be a greater impetus for migration.

The Rural Background

There are two factors relevant to a brief examination of the conditions of rural labour in UDCs: the deterioration of agricultural output and standards of living in many UDCs; and the mixed nature of the work undertaken by rural inhabitants. During the period 1950-70, food consumption and agricultural production per capita barely kept pace with population increase in UDCs. For the UDCs as a whole, food

Table 2.1: Rates of Growth of Total Agricultural Production 1955-70

	% annual rate of growth
Latin America	
1955-65	3.10
1955-70	2.74
Africa	
1955-65	2.86
1955-70	2.49
Near East	
1955-65	3.42
1955-70	3.11
Far East	
1955-65	2.90
1955-70	2.77

Source: K. Griffin, *The Political Economy of Agrarian Change* (London: Macmillan, 1975), p. 5.

production per capita rose at about 1 per cent per year, and even that modest figure was more often exceeded in the 1950s than in the 1960s. Table 2.1 shows the extent to which UDCs have experienced increases in agricultural output: the overall picture is bleak. This slow rise in standards of life is the background against which rural to urban migration has become increasingly common.

Since the input markets to which the small farmer has access will typically be imperfect, and the rate of interest charged to small farmers will be higher than that charged to bigger landholders, there is little room for the small-holder to borrow to expand his future output.[9] And even if inputs were more readily available, a risk-averting peasant, who prefers a fairly certain small income to the possibility of a larger income, might not expand his output.[10] This constraint is particularly apparent in areas where 'Green Revolution' grain varieties, requiring more sophisticated co-operant inputs such as fertiliser and irrigation, have become available. It has been argued that most of those who have been able to take advantage of Green Revolution higher yields are big landholders.

In all, some 580 million people in UDCs are living in rural poverty (defined to be an annual income equivalent to $ 50 or less). About

Table 2.2: Rural Population and Rural Poverty in Developing
Countries

Region	Rural pop. 1969	Rural population in poverty			% of rural poor in rural population		
		Incomes below $50 pa	$75 pa	One third nat. av. or $50 pa	Incomes below $50 pa	$75 pa	One third nat. av. or $50 pa
		(millions)			(percentages)		
Developing countries in:							
Africa	280	105	140	115	38	50	41
America	120	20	30	45	17	25	38
Asia	855	355	525	370	42	61	43
Developing countries total	1,255	480	695	530	38	55	42
Four Asian countries	625	295	435	295	47	70	47
Other countries	630	185	260	235	29	41	37
Share of developing countries in:							
Africa	22	22	20	22			
America	10	4	4	8			
Asia	68	74	76	70			
Total share of four Asian countries	50	62	63	56			

Source: IBRD, *Rural Development* (Washington: IBRD Sector Policy Paper,
1975), p. 80.

two thirds of these people are concentrated in four countries: India,
Indonesia, Bangladesh and Pakistan. Between regions, they are
distributed as in Tables 2.2 and 2.3.

Demand for Labour

Demand for labour-time in tropical agriculture is not constant
throughout the year: it varies seasonally by as much as 50 per cent.
To meet peaks in labour demand, extra hands will be hired, or
relatives and neighbours may be called in, in return for a feast or
other reward.[11] The seasonality of work is very strong in Egypt,

Table 2.3: Estimated Rural Population in Poverty, by Region and
Income Level of Country, 1974

Region	Rural population in poverty in countries with incomes up to $200 per capita	Rural population in poverty in other developing countries	Total rural population in poverty
Eastern Africa	60	—	60
Western Africa	15	35	50
East Asia and Pacific	10	105	115
South Asia	270	—	270
Europe, Middle East and North Africa	5	30	35
Latin America and Caribbean	—	50	50
TOTAL	360	220	580

Source: IBRD, *Rural Development* (Washington DC: IBRD Sector Policy Paper, 1975), p. 89.

especially among women and children, where small farmers rely on income from this source.[12] Up to 10 per cent of their work can be in non-agricultural labour outside their own farm; and this proportion is likely to increase with diminishing size of the home-farm.[13] In the Gambia where labour-use is extremely seasonal on farms, for six months out of the year farm-workers tend to look to tourism and other urban sources of income. There is a close relationship, in Southern Nigeria, between the rate of rural-urban migration and expenditure on hired labour on the part of those left behind: naturally as the able-bodied young move to cities, the old and children are increasingly unable to meet peak demands for labour-input with their own efforts. In Sudan, some 250,000 cotton-pickers travel up to 3,300 kilometres each year to pick at the Gezira Scheme,[14] and in Northern Nigeria farmers spend over 30 per cent of their time during the peak month in off-farm employment.[15] Similarly, 41 per cent of the males surveyed in rural Western Nigeria were engaged either entirely or part-time in non-farm activities.[16] In the case of pastoralists,

. . . the general model which regards pastoralists as being peculiarly restricted in their means of earning their livelihood just to the tending of their herds and to the consumption, directly or through exchange, of the products of their herds, . . . is not a very useful one . . . In ecologically suitable areas . . . they do, or did, obtain their livelihood from raiding, slave-trading, levying protection money from caravans, or 'rent' from oasis-dwellers, from fishing, honey- or incense-gathering, transporting goods, smuggling, making carpets or other artifacts, serving in military forces, working in mines or oil fields within the rangelands, or selling their labour for part of the year outside the rangelands.[17]

An apposite summary is that 'this fluidity of labour between a number of activities on a seasonal basis is a striking feature of rural Africa'.[18] And as for India, the many patterns of labour-use found reflect the fact that most rural workers there combine self-employment and wage employment.[19] Most of the movement which goes on is between rural areas at different times of the year. Because a lot of labour both hires and is hired out, there is naturally a great deal of migration between farms and regions. Indeed most Indian migration is rural to rural. In the Indian regions of Madhya Pradesh and Uttar Pradesh, civil construction work is arranged during the dry season of November to May, to allow small farmers (both men and women) as well as landless labourers to join in. The farmers amongst these seasonal labourers still see agriculture as their principal occupation, and travel hundreds of miles to take this work.[20] Having recognised that strict agricultural/ non-agricultural and rural/urban distinctions do not accurately reflect patterns of work in UDCs as a whole, this is a suitable place to begin a discussion of circular and seasonal migration. The needs of farm and off-farm rural labour, as has been pointed out, vary throughout the year, and it is likely that in the slack periods some extra earnings will be sought. As was related for the case of India, some people respond by making lengthy journeys between rural points to acquire extra cash income. In the cases of African workers this often takes the form of travelling to towns and cities.

Circular and Seasonal Migration

Circular migration is primarily an African phenomenon; in Latin America it is less common to return to the original place of residence. This reveals, of course, the ambiguities involved in divorcing circular migration from other categories. For while the migrant may always tend to return to his 'home', his reasons for leaving there in the first place may be no different from those affecting migrants who end up staying away permanently. The reader must beware of being overly rigid in the application of these categories of migration: they are merely convenient categories with which to begin a discussion.

The material on circular migration will be collected around two themes: the target worker effect, and the status effect. The target worker effect concerns a worker who stays in town just long enough to accumulate the cash sum (the target) which is required. Changes in the wage rate will then directly affect the number of hours worked. Figure 2.1 shows how, with a target income of OA units, fewer hours are worked (OC instead of OB) as the hourly pay-rate increases from OW_2 to OW_1.

Figure 2.1

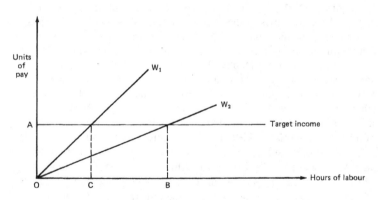

The common explanation for behaviour of this sort concerns the imposition of taxes in the colonial period. Whereas in West Africa a surplus of labour existed in a number of cities by World War I, excess demand for labour persisted into the 1950s in East Africa. Weeks attributes this to the greater population density in the rural areas of West Africa, and the high demand for labour to work on the large white-owned farms in East Africa.[21] In order to secure a sufficient effort in the export industries, poll taxes were introduced, and these had to be paid in cash, necessitating a visit to the city.[22] A variety of this in Zambia turned on the need to secure copper-mine workers. While tribal chiefs made little contribution to urban labour recruitment, taxes were used as stimulants.[23] Naturally, these explanations rely on the notion that an African worker has a finite desire to acquire manufactured goods in cities, and that, having bought these goods, he will return content to the village. But as more and more consumer goods became available, this explanation became less compelling. The need to provide wedding clothes for the bride, for instance, which was documented by Shapera,[24] is not likely to be important now; and it may be that circular migration of this 'target' variety will increasingly disappear.

Dealing now with the status effect, Caldwell quotes the reactions of a young Ashanti boy visiting town for the first time: 'I became a sort of idiot as we moved along, for I stood to gaze at whatever English-made articles I had never seen before.'[25] This perspective on the newcomer's reactions hints at the importance of visiting town in terms of the status to be gained. Shapera spoke of initiation rites;[26] and Van Velsen of the stability provided in Tongan society by this process of circular migration.[27] But these explanations can be contrasted with a simpler economic view, which is that wages and living conditions in towns were simply unattractive to young (especially unmarried) people. Only with the improvement of urban real wages in the 1950s have people been willing to contemplate a longer spell of work and residence in town. An early study of Nigerian labour turnover by Kilby found evidence that bad supervision and conditions could account for a large portion of the high turnover of workers from rural areas.[28]

Associated with this pattern of mixing urban and rural work is the observation, made in many countries, that African migrants frequently do not see themselves as permanent urbanites. One comes across phrases such as 'our sons abroad' referring to members of a rural family who are working in a town: they may see themselves and be seen as *in* the town, but they are rarely *of* the town. The urban worker who continues to feel allegiance to an ethnic or home-area association is referred to as a 'cross pressured worker'. This ambivalence to urban commitment is particularly apparent in cases where industrial workers fail to demonstrate a clear preference for trade union interests over those of their *association d'originaires*.[29] Another term for this phenomenon is for the worker to be said to live in a 'dual system'.[30]

Three aspects of this lifestyle of having a foot in two regions and occupations are of interest. The first is that it fragments the family. Industrial workers frequently have their wives, not in the town where they work, but in the home-town or farm. In Southern Africa this is best explained by very poor urban living conditions, State licensing systems, and lack of employment opportunities for spouses and children; elsewhere it more probably reflects the seasonal possibilities of farming. Statistics which have shown a recent diminution in the large preponderance of males found in cities may, in part at least, be interpreted as being due to the influx of higher-income families. A second aspect of this incomplete urban commitment is that housing behaviour reflects, not the desire to invest in urban homes, but a preference for investing any urban savings which may accrue in a rural home. And finally, and related to this, there is a flow of remittances from city to country.[31] The implication of these remittances of most interest to the study of labour markets is that it enables (and at the same time requires, since able-bodied farm labour is absent while earning the money) extra farm labour to be hired at certain times of the year. Strong relationships have been found to exist in Southern Nigeria, for instance, between local rates of labour loss through migration and expenditure in that area upon hired labour.[32]

Rural to Urban Migration

The variety of migration usually evoked by the phrases 'urban drift' and 'urban hypertrophy' is rural to urban migration. The shift to the towns and cities in UDCs, quantified in chapter 1, which has been called 'the largest migratory movement in human history',[33] will now be examined from the point of view of theoretical model-building. Trying to explain patterns of migration in so many different countries is naturally an ambitious task; and, as the evidence to be reviewed will show, some countries require different weights to be accorded to different factors. The best-known attempt to date is associated with Michael Todaro, whose two-sector migration model can be seen as an extension of the migration mechanism at work in the Lewis model.[34] In the Lewis schema, people migrate from the countryside to the city in response to assured urban employment, and without a real wage differential. In Todaro's model, however, these two parameters become variables: both the (estimated) real wage or income differential and the (estimated) probability of finding a job determine the individual's migration decision. Thus, an individual's decision to migrate is a function of the income gain of an urban job

weighted by the likelihood of finding such a job. To economists who
are interested solely in the predictive strength of a theory, the only
legitimate test of the Todaro hypothesis is to compare its predictions
with empirical results. A selection of such results will be presented
shortly, but before this is done three qualifications which have been
suggested to the Todaro model must be mentioned.

The first has been put forward by Gugler.[35] He believes that the
probability approach to urban migration is only useful if presented
in some time frame. Thus, the probability which is of concern is the
probability of securing a job within a specific time. This is a useful
modification to Todaro's equation, since it enables one to take into
account two observed phenomena: first, the likelihood that failure to
find a suitable job will result in the individual's returning to his or
her place of origin; and secondly, the individual's likely turn to
searching for an 'informal' sector job, if he or she first looked
unsuccessfully in the 'formal' sector job market. How far Todaro's
hypothesis can take into account the heterogenous mixture of
street-hawkers, pedlars, small-scale manufacturers and service
purveyors that is usually known as the 'informal' sector is a subject
of debate: this is the second qualification to his specification of the
migration process. Either the model must be framed to account only
for 'formal' sector jobs, in which case enumeration and testing are
made relatively easier (since these jobs are normally quantifiable and
earn an identifiable reward); or it must embrace jobs in the 'informal'
sector too. In the latter case, testing is made virtually impossible,
because counting those with 'informal' jobs is physically cumbersome,
and because 'jobs' in these occupations may not be full-time, or may
not entail any cash transactions, payment being in kind (for example,
food or a place to sleep).

A third difficulty with a purely predictive test of the hypothesis
is that a good statistical relationship (measured by R^2, the square of
the correlation coefficient) may be spurious. The relationship
between migration and Todaro's two variables may be close, but not
causal, the true causal factor or factors only being a proxy for one of
the two variables Todaro *has* included. This is not a pedantic point:
it is essential for policy purposes to know which factors help to
bring about migration. In other words, one needs to be able to
explain as well as simply *predict*. The distinction between the two
can be summarised as explanation concerning itself with mechanisms
and processes ('how?'), and prediction being concerned with whether
or not something happens.

A prime example of this statistical difficulty arises when the strength of education as a determinant of migration is tested. In a survey of Ghana, Godfrey found that migration continued in the face of declining job opportunities *and* a declining real income differential. The most important factor in migration decisions, as far as the data permitted interpretation, was education, he concluded.[36] If education is indeed an important factor in migration decisions – and most of the evidence suggests that it is – two problems are raised for the type of theorising presented by Todaro. First, educational attainment is likely to be related to both the expected earnings differential and the probability of finding an urban job. Even if there is excess supply of educated manpower in cities, moving to the city may still be rational if it at least raises the possibility of finding the type of job which in the countryside is unobtainable. In such cases it is virtually impossible, given the data likely to be available, to determine how far earnings differentials, perceived employment probabilities and education are responsible for migration. The second difficulty arises in ascertaining precisely how education is thought to be influencing migration. A number of ways in which education impinges upon migration can be distinguished.

First is the crude 'level' of education attained: having spent more years at school may in itself lead to migration. By the same token, in order to acquire any formal education other than primary, children in many countries are obliged to go to cities. A proxy for this is the enrolment ratio, which shows a markedly greater availability of post-primary education in urban areas. In Colombia, 60 per cent of rural schools offer only two primary years;[37] in Mexico, 60 per cent of children are enrolled in Mexico City, but only 12 per cent in outlying Oaxaca;[38] and in Kenya, 85 per cent of children in Central Province have primary education, but elsewhere in Kenya as few as 35 per cent do.[39] Allied to this is the tendency for universities to be sited in towns. Whether something inherent in 'formal' education induces disrespect for rural life has long been a controversy surrounding migration. In an essay arguing the case for more rurally-oriented curricula, and for greater prestige to be accorded country life and occupations, Balogh[40] argued that one could stem migration by changing the content of education. Others have seriously questioned this policy, however. Foster, referring to this as the 'vocational school fallacy', stated that the nature of the curriculum was irrelevant, and that only changing the relative availability of jobs between town and countryside would have an effect on migration rates.[41] Others have argued along similar lines,

mentioning the need to realise a return on educational costs (if any
have been incurred, other than opportunity costs) and the consequent
need to go to town.[42] A survey of the subject by Todaro concluded:

> Restructuring educational systems to minimise the inherent
> (and, in many cases, inherited) urban bias and to orient the
> curriculum more towards the real development needs of the nation
> (e.g. towards rural development) will have limited success in either
> curtailing rural-urban migration or changing students' attitudes
> in the absence of necessary fundamental changes in incentives
> *outside* the educational system.[43] (Emphasis in original.)

Another view of the stimulus education gives to migration may be
more accurate than Balogh's: this relates to the fact that in many
systems primary education is little more than a preparation for
further education: it does not in itself prepare people for a useful job.
Again there is some controversy as to how far schoolchildren do
actually adjust their aspirations in line with the likelihood of securing
a job. A survey of East African education considered that

> ... there is now extensive evidence to confirm that one of the
> main socialising effects of hierarchical and selective school systems
> having a close connection with the occupational reward structure
> is to instil in students levels of aspiration and expectation appropri-
> ate to their likely future position ... As the educational pyramid
> in most countries is sharply tapering in shape, the main socialising
> effect of education is to prepare the majority of students to accept
> relatively low status and small rewards.[44]

Econometric Tests

There are, then, a number of ways in which education might impinge
on migration. The precise mechanisms through which this relationship
is worked out are, as the above brief account shows, the subject of
some debate. Yet most econometric studies of migration do find
educational factors important one way or another in the individual's
decisions to migrate. This section will present a selection of the
studies of migration; the next will review some other aspects of
migration.

   The single most important conclusion to be gleaned from the
econometric studies is the importance they ascribe to economic factors

in the decision to migrate. This is in itself in perfect accord with classical economic theory, and the assumption therein that the rational economic unit (*homo oeconomicus*) maximises his utility and income by moving between regions of an economy. Differences in regional wages are thereby eliminated, and equilibrium is restored.

An early test of migration decisions was that of Ravenstein, whose 1885 articles confirmed that movements in the UK over 1871-81 conformed with the theory of income maximisation.[45]

An econometric test in Ghana using 1960 data for inter-regional moves showed that the level of education was an important factor in migration; but once again, however, establishing the precise nature of causality was hard.[46] Distance was also found to be an important determinant of migration, but income differentials were the most significant. The importance of education in Ghanaian migration was again suggested by Godfrey, whose 1973 test found the variables included in the Todaro mechanism to be inadequate.[47] (The same writer has also criticised the Todaro schema on the novel grounds that it is overly individualistic. Rather than seeing migration in UDCs as an aggregated set of individual micro-decisions, he argues that it can better be tested and understood as a macro or structural shift reflecting the changing employment pattern of a country.[48]) In Sudan, the primary reasons for migration appear to be the anticipated income differentials. In a sample survey of migrants to Khartoum, 75 per cent of the respondents said they had moved in response to the economic variables of income and employment. By 1971, in the three major Sudanese cities, 32 per cent of the population were migrants.[49] Nigerian evidence suggests over-populated lands have stimulated labour movement, especially to cocoa-producing areas with labour shortages.[50] A survey of unemployed people in Lagos ascertained that 44 per cent moved in response to job prospects, and 14 per cent to go to school.[51] Further Nigerian work found, using inter-regional data, a negative effect for income differentials and a positive effect of education.[52] The Latin American evidence is in accord with the basic conclusions from Africa: 'the similarity, broadly interpreted, of their results is testimony to the robust qualities of this type of model'.[53] Sahota's study of Brazil typically found income differentials to be a strong impetus to movement, and that distance was a strong deterrent.[54] Flinn and Cartano's Peru study usefully compares motives for both rural to rural, and rural to urban movements.[55] The results are summarised in Table 2.4. In a further study, Flinn observed that urban arrivals typically had come from smaller towns, not straight from

Table 2.4: Motives for Migration, Peru

Motive	To urban shanty (%)	To rural area (%)
To find land	0	14
To set up a business	5	11
To find a job	30	6
To escape violence	13	44
To meet relatives	17	3

Source: Flinn and Cartano.

the countryside. Forty-two per cent of his sample of a Bogota shanty town had been born in places with a population of over 2,000, and 54 per cent stated that their last place of residence was bigger than 2,000 people.[56] In Colombia[57] and Peru,[58] studies indicate the primacy of economic factors in migration (although the rural violence in Colombia in the 1950s played an important part in stimulating urban migration). Similarly, in India and the Philippines, studies give weight to the economic basis of migration.[59] Inter-regional labour movements in India also appear to be a response to opportunities for cash income, as was mentioned earlier in the chapter.

Residual Factors

It is clear from the foregoing that while migration can be 'explained' (or at least statistically accounted for) to some extent by economic factors, there are many residuals to be taken into account, too. In part these residuals must represent errors in the specification of the variables, and the crudeness of much of the data used in testing. But it is likely that certain additional factors are important in determining movements within UDCs. An earlier part of this chapter explored return migration, in its seasonal and temporary guises, and this type of movement is obviously not dealt with satisfactorily in a one-way migration model. Other factors which might influence movements, however, are the distribution of food aid, rural violence and oppression, and a preference to be with the family. These will be dealt with now before a final aspect of movement — 'step migration' — is considered.

Starvation and unreliable distribution of food (e.g. in food aid schemes) will naturally affect migration. The Sahelian drought of 1974, for example, which killed hundreds of thousands of cattle, caused many thousands of people from the countries bordering the Southern Sahara (these countries are among the world's poorest) to move south. Some made their way to the Atlantic coast, and now roam destitute in South-West Nigeria. The preference given to urban interests in the allocation of food aid (this preference being one reflection of so-called 'urban bias') has been cited as a mechanism whereby urban interests are strengthened and the poorer rural masses impoverished.[60] The desire to escape from violence such as civil war or intolerable tyranny is also likely to result in migration, irrespective of other economic variables being auspicious. A virtual civil war in Colombia from 1948 to 1958 resulted in many refugees fleeing to Bogota.[61] Figures are hard to obtain, but the dip in construction wages within Bogota can be taken as a proxy for excess supply of unskilled labour during the war years. A similar experience was the rural fighting in Indonesia from 1959 to 1961. Another aspect of rural life which may lead to migration is an oppressive gerontocracy — the system of social and economic control exerted by the family and village elders. This has been cited as an influence on young people leaving rural areas of the Ivory Coast, but is again difficult to assess since it corresponds with economic factors which can be more easily quantified.[62] Some characteristics of rural organisation may, on the other hand, also stimulate return migration. If one's claim to land lapses after a long period, there is a clear incentive to return. Age-graded social roles may also make the return attractive.[63]

The expressed desire of many migrants to be with their family or friends represents a further difficulty in the analysis of migration. Can this desire be seen as a truly independent force, leading to urban migration under many different circumstances, or is it merely a mechanism to determine which town or city is to be visited, once the fundamental migration decision has already been taken using other criteria? There is a lot of evidence regarding the reception of newcomers to towns: ethnic associations and other institutions provide various 'cushions' and adjustment facilities, and these will be dealt with in chapter 3. At the moment, however, it is important to note that the desire to be with family or friends will be a strong influence on the city or town selected. 'The role played in the process of migration by social and family contacts probably reinforces the tendency for rural-urban migration to focus on the capital; the bigger the city, the greater the chance that a potential migrant will know someone

Table 2.5: 'Primate' Cities in Selected African Countries

Country	Total pop. in millions	Primate city	Pop. in thousands	Year	% of urban pop.
Nigeria	61.45	Lagos	665	1963	19.4
Sudan	14.35	Khartoum	114	1956	17.5
Tanzania	12.23	Dar-es-Salaam	273	1967	35.8
Ghana	8.14	Accra	338	1960	21.1
Uganda	7.93	Kampala	123	1959	56.9
Malawi	4.13	Blantyre-Limbe	108	1966	69.7
Zambia	3.95	Lusaka	152	1966	19.6
Senegal	3.67	Dakar	374	1960	50.1
Chad	3.41	Fort Lamy	100	1963	38.0
Sierra Leone	2.44	Freetown	163	1963	55.8
Liberia	2.11	Monrovia	81	1962	64.8
Congo (B)	0.86	Brazzaville	200	1968	59.5

Source: L.R. Vagale, 'Metropolitan Cities of Africa: The Challenge and the Opportunity for Planning and Development' (Course Paper No. 10, Ibadan Polytechnic, 1976), p. 12.

there and choose it as his destination.'[64] This phenomenon – selecting a large city because of the greater probability of having contacts there – has given rise to concern over the appearance of very large 'primate' cities in the midst of much smaller towns. This tendency is greater in Africa than elsewhere, giving rise to the figures shown in Table 2.5. Burma, Ceylon, Malaysia, Nepal, Korea and Thailand exhibit urban primacy as well.

A notable tendency in Latin America is for migration to proceed in a series of 'steps', whereby people do not move from a hamlet or farm straight to a major city, but instead gravitate, through a graduation of larger and larger towns, to their final city.[65] This 'step migration' or 'floating migration' is evidenced by the data in Table 2.6. Support is also provided by Flinn's survey of migrants to Bogota, which ascertained that 68 per cent had travelled less than 100 miles in their final move to Bogota itself.[66] Again, McGreevey's study of Bogota found that only 50 per cent of the residents surveyed had moved there directly from their place of birth.[67] Gregory also found this in Upper Volta.[68] There is, interestingly, a close similarity between this step migration in UDCs and that of the UK in the nineteenth century. Far from moving straight from the countryside to the major centres like Birmingham, Manchester and London, research has established that people moved through a series of progressively larger towns.[69]

Table 2.6: Distribution of Migration Flows to and between Urban Areas

Country/Migration definition	Previous residence of urban migrants (%)		
	Total	Other urban	Rural
Brazil Total population in urban areas (administratively defined) who had moved within *one* year of the 1970 census count	100	76	24
Ghana Adults in urban areas (5,000 and over) who had moved within two years of the 1960 census count	100	38	62
Korea Total population in urban areas (50,000 and more) who had moved in 1965-70 period	100	26	74
India Total population in urban areas (5,000 and more) who had moved within *one* year of the 1961 census count	100	37	63

Source: IBRD, 'Internal Migration in Less Developed Countries: A Survey of the Literature' (World Bank Staff Working Paper No. 215, 1975), p. 9.

These comments have rounded out the picture presented before in purely economic terms. This brief introduction to urban migration has only touched the surface of a huge subject. Yet its inclusion in the book is necessary as a backdrop against which urban labour-market behaviour can be better studied. Some of the themes introduced in this chapter will reappear later: the continuance of rural ties in chapter 3; the consequences of rural-urban migration on urban employment creation in chapters 4, 5 and 6. Chapter 3 now turns to examine the reception of these urban migrants: the ways in which they

are assimilated to the city – indeed, whether they are assimilated
at all – and the ways in which they attempt to gain subsistence once
they arrive.

Notes

1.   H. Chenery and M. Syrquin, *Patterns of Development, 1950–1970*
      (Oxford: Oxford University Press for IBRD, 1975), p. 56.
2.   K. Davis, 'Population', in *Technology and Economic Development*
      (Harmondsworth: Penguin, 1965), p. 49.
3.   R. Farley, *The Economics of Latin America* (New York: Harper and
      Row, 1972), p. 21.
4.   Davis, p. 47.
5.   J.C. Caldwell, 'The Economic Rationality of High Fertility: An
      Investigation Illustrated with Nigerian Survey Data', *Population Studies*,
      Vol. 31, No. 1, March 1977, pp. 5-28.
6.   T.G. McGee, 'Hawkers and Hookers: Making Out in the Third World
      City: Some South-East Asian Examples', *Manpower and Unemployment
      Research*, Vol. 9, No. 1, April 1976, p. 12.
7.   F.L. Mott and O.J. Fapohunda, 'The Population of Nigeria' (University of
      Lagos, Human Resources Research Unit, Research Monograph No. 3,
      1975), p. 28.
8.   M.P. Todaro, 'Migration and Fertility', in 'Investment in Education:
      National Strategy Options for Developing Countries' (Washington DC:
      World Bank Staff Working Paper No. 196, 1975), pp. 24-30.
9.   K. Griffin, *The Political Economy of Agrarian Change* (London:
      Macmillan, 1974), pp. 28-9.
10. J.H. Cleave, *African Farmers: Labour Use in the Development of Small-
      holder Agriculture* (New York: Praeger, 1974), p. 200; S.R. Ortiz,
      Uncertainties in Peasant Farming (London University: The Athlone Press,
      1973), pp. 185-6; G.O.I. Abalu, 'A Note on Crop Mixtures under
      Indigenous Conditions in Northern Nigeria', *Journal of Development
      Studies*, Vol. 12, No. 3, April 1976, pp. 212-20.
11. G. Macpherson and D. Jackson, 'Village Technology for Rural Develop-
      ment', *International Labour Review*, Vol. 111, No. 2, February 1975,
      p. 99.
12. B. Hansen and M. El Tomy, 'The Seasonal Employment Profile in
      Egyptian Agriculture', *Journal of Development Studies*, Vol. 1, 1965,
      pp. 399-409.
13. B. Hansen, 'Employment and Wages in Rural Egypt', *American Economic
      Review*, Vol. 59, No. 3, 1969, pp. 298-313.
14. M. El Mustada Mustafa, 'The Sudanese Labour Market: An Overview of its
      Characteristics and Problems with special emphasis on the Urban Labour
      Market', *Manpower and Unemployment Research*, Vol. 9, No. 2,
      November 1976, p. 42.
15. D. Byerlee and C.K. Eicher, 'Rural Employment, Migration and Economic
      Development: Theoretical Issues and Empirical Evidence from Africa'
      (East Lansing: Michigan State University African Rural Employment
      Paper No. 1, 1972), p. 10.
16. C. Liedholm, 'Research on Employment in the Rural Non-farm Sector in
      Africa' (East Lansing: Michigan State University African Rural

Employment Paper No. 5, 1973), pp. 3-5.

17.   S. Sandford, 'Pastoralism Under Pressure', *ODI Review*, No. 2, 1976, p. 65.

18.   Liedholm, p. 4.

19.   K. Bardhan, 'Rural Employment, Wages and Labour Markets in India', *Economic and Political Weekly* (Bombay), 25 June 1977, pp. 34-48.

20.   IBRD, 'Some Aspects of Unskilled Labour Markets for Civil Construction in India: Observations based on Field Investigation' (Washington DC: World Bank Staff Working Paper No. 223, 1975), p. 13.

21.   J.F. Weeks, 'Wage Policy and the Colonial Legacy: A Comparative Study', *Journal of Modern African Studies*, Vol. 9, No. 3, October 1971, pp. 361-87.

22.   W. Elkan, 'Circular Migration and the Growth of Towns in East Africa', *International Labour Review*, Vol. 96, No. 6, 1967, pp. 58-89.

23.   H. Heisler, *Urbanization and the Government of Migration: The Interrelations of Urban Life in Zambia* (New York: St Martin's Press, 1974), pp. 30-45.

24.   I. Shapera, *Married Life in an African Tribe* (London: Faber and Faber, 1940).

25.   J.C. Caldwell, *African Rural-Urban Migration* (Canberra: Australian National University Press, 1969), pp. 9, 10.

26.   Shapera, op. cit.

27.   J. Van Velsen, 'Labour Migration as a Positive Factor in the Continuity of Tonga Tribal Society', *Economic Development and Cultural Change*, Vol. 8, No. 3, April 1960, pp. 265-78.

28.   P. Kilby, 'African Labour Productivity Reconsidered', *Economic Journal*, Vol. 71, No. 282, June 1961, pp. 273-91; E.J. Berg, 'Backward Sloping Labour Supply Functions in Dual Economies', *Quarterly Journal of Economics*, Vol. 75, 1961, p. 468.

29.   R. Melson, 'Ideology and Inconsistency: The "Cross-pressured" Nigerian Worker', in R. Melson and H. Wolpe (eds.), *Nigeria: Modernization and the Politics of Communalism* (Michigan: Michigan State University Press, 1971), pp. 581-605.

30.   J. Gugler, 'Life in a Dual System: Eastern Nigerians in Town, 1961', *Cahiers d'Etudes Africaines*, Vol. 11, No. 3, 1971, pp. 400-21.

31.   A.R. Waters, 'Migration, Remittances and the Cash Constraint in African Smallholder Economic Development', *Oxford Economic Papers*, Vol. 25, 1973, pp. 437-54; G.E. Johnston and W.E. Whitelaw, 'Urban-Rural Income Transfers in Kenya: An Estimated Remittances Function', *Economic Development and Cultural Change*, Vol. 22, No. 3, April 1976, pp. 473-9; W. Elkan, 'Is a Proletariat Emerging in Nairobi?', *Economic Development and Cultural Change*, Vol. 24, No. 4, July 1976, pp. 695-706.

32.   S.M. Essang and A.F. Mabawonku, 'Impact of Urban Migration on Rural Development: Theoretical Considerations and Empirical Evidence from Southern Nigeria', *The Developing Economies*, Vol. 13, No. 2, June 1975, pp. 137-49.

33.   O.H. Koenigsberger, 'The Absorption of Newcomers in the Cities of the Third World', *ODI Review*, No. 1, 1976, p. 59.

34.   M.P. Todaro, 'A Model of Labour Migration and Urban Development in Less Developed Countries', *American Economic Review*, Vol. 59, No. 1, 1969, pp. 138-48.

35.   J. Gugler, 'A Further Note on Unemployment Rates in Developing Countries', *Manpower and Unemployment Research in Africa*, Vol. 4, No. 1, April 1971, pp. 14-16.

36. E.M. Godfrey, 'Economic Variables and Rural-Urban Migration: Some
     Thoughts on the Todaro Hypothesis', *Journal of Development Studies*,
     Vol. 10, No. 1, October 1973, pp. 66-78.
37. ILO, *Towards Full Employment: A Programme for Colombia* (Geneva:
     ILO, 1970), p. 219.
38. F.H. Harbison, J. Maruhnic and P. Resnick, *Quantitative Analyses of
     Modernization and Development* (Princeton: Princeton University
     Industrial Relations Section, 1970), Appendix VIII.B.
39. ILO, *Employment, Incomes and Equality: A Strategy for Increasing
     Productive Employment in Kenya* (Geneva: ILO, 1972), p. 511.
40. T, Balogh, in J.W. Hanson and C.S. Brembeck (eds.), *Education and the
     Development of Nations* (New York: Holt, Rinehart and Winston,
     1966).
41. P.J. Foster, 'The Vocational School Fallacy', in Hanson and Brembeck.
42. A.R. Jolly, 'Rural-Urban Migration: Dimensions, Causes, Issues and
     Policies', in R. Robinson and P. Johnston (eds.), *Prospects for Employment
     Opportunities in the 1970s* (London: HMSO, 1971), p. 123.
43. M.P. Todaro, 'Migration and Fertility', p. 26.
44. D. Court, 'The Education System as a Response to Inequality in Tanzania
     and Kenya', *Journal of Modern African Studies*, Vol. 14, No. 4, December
     1976, p. 664.
45. E.G. Ravenstein, 'The Laws of Migration', *Journal of the Royal
     Statistical Society*, Vol. 48, No. 2, June 1885, pp. 167-227.
46. R. Beals, M. Levy and L. Moses, 'Rationality and Migration in Ghana',
     Review of Economics and Statistics, Vol. 49, No. 4, November 1967,
     pp. 480-6.
47. E.M. Godfrey, op. cit.
48. E.M. Godfrey, 'A Note on Rural-Urban Migration: An Alternative
     Framework of Analysis', *Manpower and Unemployment Research*, Vol. 8,
     No. 2, November 1975, pp. 9-12.
49. M. El Mustada Mustafa, p. 38.
50. A. Adepoju, 'Migration and Development in Nigeria', *Manpower and
     Unemployment Research*, Vol. 9, No. 2, November 1976, pp. 65-76.
51. O.J. Fapohunda, 'The Characteristics of the Unemployed People in Lagos'
     (University of Lagos, Human Resources Research Unit, Research Paper
     No. 4, 1975), p. 11.
52. A. Mabogunje, 'Migration Policy and Regional Development in Nigeria',
     Nigerian Journal of Economic and Social Studies, Vol. 12, No. 2, 1970,
     pp. 243-62.
53. IBRD, 'Some Economic Interpretations of Case Studies of Urban Migra-
     tion in Developing Countries' (Washington DC: World Bank Staff Working
     Paper No. 151, 1973), p. 6.
54. G. Sahota, 'An Economic Analysis of Internal Migration in Brazil',
     Journal of Political Economy, Vol. 76, No. 2, March-April 1968, pp.
     218-45.
55. W.L. Flinn and D.G. Cartano, 'A Comparison of the Migration Process to
     an Urban Barrio and to a Rural Community: Two Case Studies', *Inter-
     American Economic Affairs*, Vol. 24, No. 2, Autumn 1970, pp. 37-48.
56. W.L. Flinn, 'The Process of Migration to a Shanty-Town in Bogota,
     Colombia', *Inter-American Economic Affairs*, Vol. 22, No. 2, Autumn
     1968, pp. 77-88.
57. R.A. Berry, 'Open Unemployment as a Social Problem in Urban Colombia:
     Myth and Reality', *Economic Development and Cultural Change*, Vol. 23,
     No. 2, January 1975, pp. 276-91.

58.   C.D. Scott, 'Peasants, Proletarianisation and the Articulation of Modes of Production: The Case of Sugar-cane Cutters in Northern Peru, 1940–1968', *Journal of Peasant Studies*, Vol. 3, No. 3, April 1976, pp. 321-42.
59.   H. Lubell, *Calcutta: Its Urban Development and Employment Prospects* (Geneva: ILO, 1974); H.T. Oshima, 'Labour-Force "Explosion" and the Labour-Intensive Sector in Asian Growth', *Economic Development and Cultural Change*, Vol. 19, No. 2, January 1971, pp. 161-83.
60.   M. Lipton, 'Urban Bias and Food Policy for Poor Countries', *Food Policy*, Vol. 1, No. 1, November 1975, pp. 41-52.
61.   A.T. Udall, 'The Effects of Rapid Increases in Labour Supply on Service Employment in Developing Countries', *Economic Development and Cultural Change*, Vol. 24, No. 4, July 1976, pp. 765-85.
62.   H. Joshi, H. Lubell and J. Mouly, 'Urban Development and Employment in Abidjan', *International Labour Review*, Vol. 111, No. 4, April 1975, pp. 289-306.
63.   J.M. Nelson, 'Sojourners versus New Urbanites: Causes and Consequences of Temporary versus Permanent Cityward Migration in Developing Countries', *Economic Development and Cultural Change*, Vol. 24, No. 4, July 1976, pp. 721-57.
64.   Joshi *et al.*, p. 295.
65.   G. Breese, *Urbanization in Newly Developing Countries* (Englewood Cliffs, N.J.: Prentice-Hall, 1966), p. 83.
66.   Flinn, pp. 82-8.
67.   W.P. McGreevey, 'Migration and Policies for Urban Growth in Colombia', unpublished paper; 'Urban Growth in Colombia', *Journal of Inter-American Studies*, Vol. 16, No. 4, November 1974, pp. 387-408.
68.   J. Gregory, 'Migration in Upper Volta', *African Urban Notes*, Vol. 6, No. 1, 1971, pp. 44-52.
69.   T.S. Ashton, *The Industrial Revolution* (London: Hutchinson, 1948).

3 THE RECEPTION OF URBAN NEWCOMERS

The previous chapters have examined how people come to leave their farms, smallholdings or villages and begin a series of moves, possibly culminating in a move to a major city. Their various reasons were briefly recounted, and the apparent preferences for larger towns noted. This chapter will examine the processes whereby the 50.6 million people who are estimated to arrive in UDCs' cities every year until 2000[1] are absorbed. Once again, it is necessary to point out that different countries display different patterns of assimilation; in some, indeed, assimilation of newcomers may not take place in any simple sense at all. The chapter will therefore begin with a brief examination of the so-called rural-urban continuum, to assess the extent to which distinct economic and social patterns are fostered in cities. The following sections will then look at immigrant adjustment to life in shanty towns, and at the initial stages of newcomers' economic behaviour. Intra-urban movements will also be introduced.

The Rural-Urban Dichotomy

Recent writing on urbanisation in UDCs has come to question the appropriateness of a rural-urban dichotomy in which different life-styles and attitudes are supposedly held by townspeople as opposed to country people. The early tradition in urbanisation studies was to follow Wirth's classic analysis in which 'the city shows a kind and degree of heterogeneity of population which cannot wholly be accounted for by the law of large numbers',[2] and in which the dynamic of change and influence flows from city to countryside. If, however, the city is largely growing due to the immigration of small-town and rural people, it seems probable that their tastes and styles of life will come to dominate parts of the city. McGee has referred to 'pseudo-urbanization',[3] whereby the characteristics of the city are increasingly determined by new arrivals retaining their past cultures, rather than by arrivals' assimilation of some established urban culture. As Janet Abu-Lughod writes, 'numbers alone should alert us to the probability that migrants are shaping the culture of a city as much as they are adjusting to it'.[4] Research now tends to see the lack of assimilation of newcomers in terms of preference for and use of various family networks, rather than

47

denial of access to privilege in the city. Thus, rather than the city
being characterised by secondary relationships between people —
each interaction being limited to a highly factionalised part of a
person's life — kinship ties and primary relationships persist.[5]
Wirth's classic view of individuation taking place in the city is
therefore now tempered with the observed durability of kinship
relationships.

A recent and unusually thorough examination of adaption to
urban life in Rio de Janeiro by Perlman provides a lot of evidence in
favour of the newer interpretation of urbanisation.[6] She finds the
traditional view that shanty town dwellers are 'lonely and rootless
. . . unable to adapt fully to urban life, and perpetually anxious to
return to their villages' unfounded. Family breakdown, crime and
disease, the expected concomitants of this rootlessness are similarly
missing in large measure. Instead there is an ordered set of
institutions and customs which attempt to deal with jurisdictional
problems, and a strong sense of community and responsibility amongst
the *favela* dwellers.

Adjustment to the city consists of finding subsistence during the
first period of the visit, locating sources of income from full-time or
part-time jobs, and possibly finding contacts and credit sources with
which to begin in business. It is worth remembering that the majority
of those who are arriving in the cities of Latin America will have
previously lived in towns, and may be progressing, via 'step migration',
to still larger cities. This process, mentioned in chapter 2, will of
course condition the behaviour of the new arrivals.

A definition of immigrant adjustment has been provided by
Imoagene: 'those mechanisms which exist to fill the gap created for
the migrant who leaves his homogenous primary group of kinsmen and
relatives to settle in this heterogenous urban community.'[7]

A simple test of hypotheses in the Nigerian community of Sapele
established that adjustment (which was proxied by eight or so
questions) was not enhanced as length of city-stay increased, unless
the individual had joined a clan to assist him. In Eastern Nigeria,
ethnic associations, or *associations d'originaires* provide such
assistance. Gugler's study concluded that this assistance was so
indispensable for city adjustment that calling these associations
'voluntary' was misleading.[8] Imoagene's study found that the
associations helped in cases of illness or arrest (49 per cent), for loans
(44 per cent), sickness and funeral expenses (32 per cent), in return
for which fees had to be paid. A variant is documented in Jakarta,

where 'tokes' find jobs, credit and accommodation for newcomers in return for a percentage of their future income.[9]

Similar associations exist in the cities of Latin America. Screening of new inhabitants of *barriados* is carried out by a tenants' association, which is elected annually, in the case of Lima.[10] The association serves also to adjudicate land rights disputes and levy taxes to finance certain co-operative projects. In contrast to ethnic or tribally-based groups in Africa, the Peruvian associations gather together those from certain geographical regions of the backlands.[11] Apart from the attractions of the clubs in their permitting 'social continuity not only during the stressful initial period of adjustment to life in the metropolis, but for a lifetime',[12] they can be used to initiate individuals' political ambitions, and to secure contacts for use in later careers. Again, these clubs bring rural and town life into the city: as Doughty's survey of Lima concludes, 'one is impressed by the institutional resiliency of the highland migrant; his family, circle of friends, means of recreation and indeed his community are in varying degrees reconstituted and maintained in the city.'[13]

A second form of mediation in becoming a shanty town dweller is the search for work. There are many aspects to job-search in UDCs, most of them requiring the refinement of the concepts used in Western economics. These will now be presented, before going on, in later chapters, to look at some of the factors conditioning the success of job-search.

Job-Search

An important characteristic of life in shanty towns is that many of the inhabitants will spend much of their time in avid, if haphazard, searching for new, different or supplementary jobs. The fact that many of the jobs they seek are part-time, seasonal or for non-cash rewards, or may be illegal, compounds the problems caused by the inadequacy of data mentioned before. In chapter 1 it was argued that the conventional concepts of 'employment' and 'unemployment' are not particularly useful categories into which to squeeze shanty dwellers, and the same strictures apply to 'job-search'.

To build a complete picture of the process of job-search in UDCs' cities, three sets of issues must be considered. The first of these is the time taken by different sorts of aspirants to locate a suitable job. Not everybody, of course, will find work with equal ease: those possessing certain qualifications, experience or contacts, will be in greater demand. Connected with this is the ease of entry to various occupations. Some may be relatively easy to begin – street selling, perhaps; others will entail the painstaking cultivation of contacts and the accumulation of some capital. Second is the type of job people aspire to, both in the short run and the long run. People may be prepared to spend some time seeking out a job which they particularly want, rather than to accept the first vacancy they find. Although they may be more prepared to compromise in the shortrun, simply to secure an income during their initial months in the city, they are likely to have other aims in mind, towards which they hope to progress as their stay continues. Finally, one must consider the ways in which people seek out opportunities – the channels they use for information and selection. In the absence, in most UDCs, of an effective formal employment exchange system, different marketplaces and forums will spring up to enable labour-market information to be disseminated.

How long, then, will a job-seeker take to find a source of subsistence? Posed in this way, the answer is misleading, for it ignores the complexity of the issues just raised: there is a great heterogeneity of jobs and people to fill them. Despite the conceptual and empirical difficulties of talking about 'the' job-seeker, however, some evidence is presented below, in Table 3.1, before the complications are discussed. Table 3.1 indicates the time taken by various sample groups to secure a first job in Santiago, Brazilian cities, Seoul, Lima, Tanzanian cities, Buenos Aires, Rio de Janeiro and Bogota. The clear impression is that after one month at least two-thirds of job-seekers have found a job. But this treatment obviously has to be qualified with the questions raised at the beginning of this section.

First, one needs to know what proportion of these apparently successful job-seekers arrived with their jobs already arranged. These people's fortunate experience will obviously colour the data for the rest, biasing the results towards a shorter average duration of search. The practice of not going to Colombian cities without a job arranged seems widespread.[14] In Rio de Janeiro, it has been estimated that 13 per cent arrive with a job already arranged;[15] in Bombay, the data is better, and is shown in Table 3.2. It is noticeable in this table that the Hindi have many more jobs arranged in advance than either the Marathi or Tamils – this presumably reflects their tendency to maintain rather tighter-knit communities and to pass on information more assiduously. Still, 38 per cent of Marathi and higher percentages of the other two groups were prepared to travel to Bombay without a job having been arranged.

Table 3.1: Time Required for Migrants to Find First Job in City

City	Sample Description	Cumulative percentage finding work within		
City-wide samples				
Santiago, Chile	Economically active migrants who arrived in Santiago within previous decade	43 66 85	immediately (2 days) 1 month 6 months	
Brazil: 6 cities including Rio and Sao Paulo	Adult migrants	*Male* 85 95	*Female* 74 90	1 month less than 6 months
Seoul, Korea	Household heads, of whom 80 per cent are migrants	26 64 76	immediately (pre-arranged) 'soon' 6 months	
Lima, Peru	1967 survey of migrants	over 75	3 months	
Bogota, Colombia[a]	Migrants	80 25	2 months 5 weeks	
Rural-urban sample				
Tanzania: urban areas	Males who moved from rural areas to urban areas after the age of 13	80 over 90	3 months 6 months	
Poor sections of city				
Santiago, Chile	Family heads or their wives in a *callampa* settlement. 85 per cent manual labourers or self-employed artisans	47 91	'immediately' 3 months	
Buenos Aires, Argentina	Residents of a *villa miseria*, mostly recent migrants, 61 per cent day labourers or unskilled workers	74 85	2 weeks 1 month	
Rio de Janeiro, Brazil[b]	Residents of three *favelas*. (a) Those with experience in unskilled urban or rural work (b) Those with previous skilled jobs (c) Those with help in finding (d) Those with no help	85 65 43 67	1 month 1 month 1 month 1 month	

[a] R.A. Berry, p. 287.
[b] Perlman, p. 80.

Sources: IBRD, 'Internal Migration in Less Developed Countries: A Survey of the Literature' (Washington DC: World Bank Staff Working Paper No. 215, September 1975), p. 26.

Table 3.2: Job Information Among Arrivals to Bombay

	Group, %		
Arrival	Marathi	Hindi	Tamil
On appointment	28	4	9
Promised a job	5	21	9
Seen advertisement	3	3	3
Nothing in mind	38	55	61
All non-transferees	74	83	82

Source: Joshi and Joshi, p. 132.

The second major difficulty of data on speed of absorption is that it provides no consistent criteria on what sort of activity represents a job. The definitional niceties of this have been discussed already; but here, for once, it would be useful to have some unambiguous point of reference. Many of those who arrive without jobs could presumably live for a while by stealing or illegal street peddling, and this reduces one's ability to determine the interaction of supply and demand in urban labour markets.

Speed of assimilation has been documented in a few cities, and while none of the studies has completely overcome the difficulties just alluded to, they are worth reviewing.

There is near unanimity that 'educated labour' — judged by roughly comparable criteria in each study — spends longer looking for work than does uneducated labour. Some portion of overt urban unemployment, then, is to be interpreted as voluntary. Ramos's survey of Latin American employment argued in a manner similar to other studies:

Since . . . secondary-school graduates generally come from higher-income families, such workers are in a better position to discriminate among job possibilities and not simply take the first job offered to them. The result is that unemployment rates may appear to rise among workers with secondary education . . . in fact, their higher-income families may simply be permitting them to be more choosy as to the work they accept.[16]

Similarly: 'what little Colombian evidence has been adduced to date

Table 3.3: Education and Duration of Unemployment: Urban India
1961-2

Education	Duration of unemployment (%)		
	Less than 1 month	1-9 months	Over 9 months
Secondary school	9	39	52
Literate, below secondary	15	37	48
Illiterate	47	34	19
Overall average (includes other groups)	21	38	41

Source: D. Turnham and I. Jaeger, *The Employment Problem in Less Developed Countries: A Review of Evidence* (Paris: OECD, 1971), p. 52.

on how unemployment is generated is consistent with the conclusion that most urban unemployment has a 'voluntary' component. Most people who leave their jobs appear to do so by their own choice, rather than through the action of their employer.'[17] A study by Prieto on this point found that most voluntary quits in Bogota were caused by the desire to find a higher salary.[18] In Rio de Janeiro, for educated labour, 'the help of friends and relatives seems to have helped delay the timing and thereby improved the quality of employment'.[19] Thus, those who took three months or more before selecting a job got 'desirable public or private security jobs (police, guards)'. There is also supporting evidence in Todaro's overview of migration.[20] In the Indian Sample Survey of 1961-2, duration of unemployment behaved similarly with respect to educational attainment. Table 3.3 reproduces the results.

The Gambia has unusually strong seasonal changes in demand for labour, and for this reason does not provide readily generalisable experience. But the behaviour of its educated unemployed is interesting. Those with no schooling have a registered unemployment rate of 6 per cent; those with primary education 4 per cent; and those with secondary education 12 per cent.[21] Secondary school leavers spend a long time looking for work, ensuring that they maximise their intended seasonal or lifetime earnings. There is the additional argument that those who have completed secondary school are more likely to have wealthy backgrounds than those with less education,

and can thus prolong their spell of job-search.
One proxy for voluntary unemployment and job-search is the
voluntary quit rate. But here again data are lacking and their
interpretation bedevilled with the possibility that quits reflect the
desire to return home from the city and not the chance of a better
job in town. Quit rates in Kenya show a marked reluctance on the
part of the older, less educated workers, to leave a job. Rates are
commonly as low as 10 per cent, with some plants achieving no
voluntary turnover at all.[22] Mobility between jobs is low in traditional
industrial jobs in India, textiles for instance. The average duration of
employment there is eight years.[23] Desire for a better standard of
living there tends to be reflected in trying to find another family
member a job, rather than by the main breadwinner looking for a
different job. This behaviour may reflect poor labour-market
information (although the extent of this information will usually
differ by workers' level of education); alternatively, it may be
caused by the low perceived dispersion of wages obtainable between
industries.

The contradiction of this evidence with cross-sectional evidence
on rates of unemployment by level of education is clear, but resolvable.
It will be shown in chapter 4 that slow increases in demand for labour
in the cities of UDCs affect parts of the labour force differently, one
of the determinants of success being educational attainment. While,
at any given time, more-educated people may be experiencing less
difficulty than less-educated people in getting a job, they will tend to
take longer doing it. In the knowledge that they will be presented with
more opportunities than just the first they discover, and with the
greater need to maximise a return on the education they have acquired,
their search is more careful and prolonged.

The subject of how easy it is to begin working in different
occupations is explored at some length in chapter 5, where the
hypothesis that the 'informal' sector is an employer of last resort
for all labour is examined. One of the themes of this book is that the
formal-informal dichotomy, while a useful shorthand for those who
know its shortcomings, hampers the understanding of much urban
economic activity. But for the sake of this section, it will be assumed
that one can talk of a 'formal' sector. Entry to its various parts can
then be examined without the need for constant qualification.

Entry to many UDCs' large private firms, or public sector, is
increasingly policed by screening systems. To overcome these barriers,
formal educational qualifications, or work experience, or both, will,

in varying mixes, be important. If the output of a national education system outruns the economy's ability to employ it, with a given wage-structure, one or both of two consequences will follow. First, the wage applicable to the job may fall: since there is excess supply of applicants, the employer can presumably lower his offer-price and still sttract some recruits. How far he can do this is a straight-forward function of the elasticity of supply of job-seekers: some may be prepared to take the job at a substantially lower wage; others, faced with this, will look elsewhere or even leave the labour force. Alternatively, the employer may require better labout for the same wage. He may insist on a secondary school certificate rather than a primary school certificate; at a different part of the market, he may require a Ph.D. rather than the hitherto sufficient M.A.[24] Or he may do both.

In the latter case, there will be a tendency for the best-qualified people to pre-empt the best jobs — even if those 'best' jobs are unskilled. But so long as possession of ever higher qualifications is needed to gain entry to an occupation, it will be rational to continue in education until the requisite qualifications are obtained. The desire to leave the schooling system before its peak (university or secondary school, depending on different countries), already likely to be slight due to the disproportionate wage differential enjoyed by those at the top of the hierarchy, is thereby further diminished.[25]

The notion of entry being policed or screened as explained above implies the existence of 'internal labour markets'. These are to be differentiated from competitive labour markets with homogenous labour in two ways. First, people cannot enter such firms' labour forces at will. They can only enter (through 'ports') at certain stages in the hierarchy which are agreed upon by the management and trade unions.[26] Second, corresponding to each entry-port is a different level of skill or occupation, reflecting the fact that within the labour force there is a formal differentiation of grades. This system suits management because it knows that it has cadres of (usually) internally-trained labour which can be promoted to overcome any shortages on the open labour market; and it suits unions because they can more readily protect their members from arbitrary dilution by new entrants and guarantee them some measure of orderly promotion and lay-off procedures. What evidence is there on employers' preferences regarding qualifications? How likely is it, in other words, that an uneducated newcomer to a third world city will find a ready source of income in a 'formal' plant?

In Ghana, employers in private firms prefer their recruits to have had experience of self-employment.[27] In Colombian manufacturing, some schooling is sought, but no work experience is necessary. Hiring is mostly at entry-level.[28] King's short tracer-study of a Kenyan manufacturer follows him between private firms and self-employment.[29] Other studies by the same writer have indicated the highly developed apprenticeships used by Indian-owned enterprises in Kenya,[30] in which apprentices pay to be shown a craft, after having left secondary school. Still on Kenya, Fields has suggested that 'formal' employers tend to prefer recruits with formal educational credentials, if the wage premium attaching to them is not substantial.[31]

A proportion of those who arrive in towns and cities without jobs already arranged will, therefore, have their ambition thwarted by the better-educated labour against which they must compete. With varying speeds, some will then turn to less prestigious, or lucrative, or secure sources of income. Their search will then increasingly turn to the 'informal' sector.

The transition which job-seekers make from canvassing solely 'formal' establishments to considering taking 'informal' jobs as well is ill-understood and little documented. Indeed, comprehension is in large measure hindered by the conventional use of the formal-informal labour market dichotomy, as chapter 5 will argue. But a few principles of this behaviour are understood.

If there is excess demand for labour with certain types of qualification, but excess supply of labour without these sought-after qualifications, one would expect the former labour to locate jobs more quickly. It is possible, however, that these people, by virtue of their being in demand, will perceive a greater dispersion of wage offers than that facing other types of labour. To the extent that this is true, it becomes rational for the qualified labour to spend more time looking for jobs, or, in human-capital theory parlance, to invest in job-search procedure. Standard job-search models posit an aspiration level (a wage and non-wage benefit combination sought from employment) which declines, linearly or otherwise, through time as the job-seeker fails to find a job.[32]

A number of factors will determine the speed with which the aspiration of the job-seeker falls. On the supply side, the first factor will be the funds to which the job-seeker has access. Recent arrivals to cities tend not to arrive penniless and without connections, as chapter 2 indicated. In addition, migrants tend not to be the very poorest of the rural people anyway.[33] Next is the ability or willingness of friends and relatives to underwrite this process: this will eventually begin to decline.[34] Another factor, which has been referred to already in this chapter, is the level of education of the job-seeker. Fapohunda's study of the Lagos unemployed, for instance, established fairly strong differences in intent by education, as Table 3.4 shows. 'Only 22 per cent [of the sample of 1,480] of unemployed people with a school certificate, and 17 per cent of unemployed university graduates, receive any casual income'; and 'between 76 per cent and 93 per cent of any group of people who have been unemployed for a given period have a minimum education of primary school passes'.[35] Thus, not only do the better educated have strong preferences for the type of work they want — predictably enough — but they are less likely to take 'casual work' while searching for a better job. This is presumably in the belief that this distraction will not help in securing a job.

Table 3.4: Job Preference and Level of Education: Lagos, 1974

Level of education	Job sought, %		
	'Anything'	More discriminating answer	Total
No formal	54	46	100
1 to 7 years	49	51	100
7 to 12 years	28	72	100
Tertiary	25	75	100

Source: O.J. Fapohunda, 'Characteristics of the Unemployed People in Lagos' (University of Lagos Human Resources Research Unit, Research Paper No. 4, Lagos, 1975).

A final factor likely to be important in speed of adjustment of aspirations to labour demand is taste. Different people will naturally have different preferences for risk, reflecting their estimated chances of finding work of various types. In British labour economics the typology of 'stickers' who stay in one area in the hope of finding a job there eventually, and 'snatchers' who will accept an offer in a different area, has been devised.[36] The demand factor which determines speed of assimilation to the labour market is, of course, the intensity of demand for labour of different categories.

A problem which arises from the distinction which has been drawn here between searching for a job and having a job is that the two are likely to be construed as mutually exclusive. In developed countries, this is usually considered to be a reasonable distinction, and is buttressed by surveys of workers' knowledge of local labour-market conditions, which usually reveal considerable ignorance of conditions outside the plant.[37] In UDCs, on the other hand, it is likely that many people will spend part of the day looking for work even if they have some sort of a job. They may do this in anticipation of losing their current job, because they are risk-averse and want better information which they feel thay can fall back on, or because they may be underemployed in their present job. A survey of workers in three low-income areas in cities in Latin America[38] established that two-thirds of those who wanted to change their jobs were under-utilised, in the sense that their income fluctuated, typically giving them only 60 per cent of regular workers' income. To show these ideas in a simple diagram, what is being suggested is that the aspirations curve in Figure 3.1 declines after time OA spent fruitlessly searching for work.

Figure 3.1

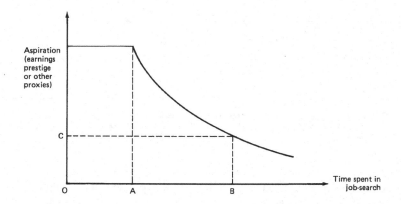

Eventually, after the lapse of further time AB, the aspirations curve may even have sunk beneath the required daily income level OC, at which point casual earnings and other 'informal' earnings become acceptable. By time OB, then, the job-seeker will have almost abandoned his intention of taking only regular 'formal' work and will be spending most of the day in other types of work. But there is a further complication. The job preferences explained above cannot be seen, in UDCs' cities, as once-and-for-all choices which remain unchanged throughout individuals' working experience. On the contrary, it is likely that preferences between occupations, between wage-employment and self-employment and between mixes of the above, will themselves change as individuals' perceived options open and close. Experience may be important for entry into factories in some countries, so that self-employment or some variety of 'informal' work will be taken before wage-employment is a feasible aim. Alternatively, a spell of wage-employment is likely to be used in some countries as an investment for building up patronage networks for use in a later career as an independent operator. Chapter 5 will elaborate upon this point, since it is an important strain of criticism in the 'informal' sector debate. Here, it is important simply to note that 'formal' and 'informal' jobs are not pursued exclusive of one another.

How Jobs Are Found

The information channels used by job-seekers will reflect the absence of effective state institutions to collect information. In chapter 2 it was noted how often migration took place to towns or cities in which friends and relatives live; and earlier in this chapter, the role of these people, and the institutions which formalise the assistance they provide the urban newcomer, in helping find work, was mentioned. Some more details of this are now presented.

Of migrants to Bombay, 53 per cent of the Marathi, and 71 per cent of both Hindi and Tamils have relatives already there.[39] A form of information provision common to Bombay is villages' tendency to concentrate on filling certain occupations. Thus, one village will habitually provide some of the Bombay police force; another cotton-mill workers, and so on.[40] The return-migration at the end of the stay in town of course strengthens the links between generations of workers in a village, and their eventual urban employers. Perlman's study of Rio de Janeiro determined that 45 per cent of migrants did not begin to look for work straight after arrival; and, when they did begin, two-thirds found their first work through the efforts of friends.[41] In Almedabad, India, friends, relatives and neighbours provided 68 per cent of first time job-seekers with their information on job openings; while advertisements were only used by 20 per cent, and employment exchanges by 1.8 per cent.[42] In a Colombian survey, 40 per cent of job-seekers received help from friends in finding their first job;[43] in Lagos, 30 per cent had to resort to bribery to secure a job;[44] in Kenya, use of patronage was reported as extensive;[45] in Ghana, 50 per cent of job-seekers used family contacts;[46] and in Calcutta, 56 per cent of job-seekers used similar contacts.[47] Employment exchanges are little used by unskilled job-seekers in India. Seven per cent of large private and public employers recruit from employment exchanges in Bombay; 60 per cent to 65 per cent use 'direct' or personal and patronage recruitment methods.[48] This is despite one's ability to register at more than one exchange. On rural sites, among Indian migrant workers, a certain proportion of the wage has to be given to gang-leaders, who act as brokers with the employers.[49]

Different considerations apply to job-changers, however. In this case, in Almedabad, the use of advertisements rose to 40 per cent, and the use of friends and other workers fell to 40 per cent. This is to be expected, since job-changers are likely to have been in the labour market looking for work more recently, and possibly more often, than those who never change their jobs.[50]

Finally, moving from the consideration of individuals' job-search and choice to a view of a family unit's operation, some more modifications of perspective are required. There is evidence that a large degree of intra-family or intra-community job-sharing takes place in shanty towns. This allows greater flexibility of working hours, or of movement. If the need arises for one member of the family to return home suddenly, a substitute source of hours of work can readily be found. For this reason, it is extremely difficult to investigate 'informal sector households', when incomes are earned by substitution of jobs between members of the family, and where there may be extensive cross-subsidisation of working capital to initiate new businesses. The problem is well put by Bienefeld and Godfrey:

. . . if we are primarily interested in welfare, the object of statistical investigation must be the household, since this is the primary income-sharing mechanism, and the operative distinction has to be one based on poverty levels or income levels. This means, of course, dropping the formal/informal sector distinction, since individuals within the same household may work in different sectors of the economy.[51]

A rare survey of income-sharing is based on Senegal, where it was found that only 31 per cent of a sample of casual industrial workers (most of them labourers) relied solely on their casual pay for survival. About 45 per cent of their additional incomes came from farm labour, about 25 per cent from trading and the rest from a miscellany of other sources.[52] Similar observations apply to Rio de Janeiro, where many of those sampled in various *favelas* were found to hold more than one 'job' at a time.[53]

The above discussion shows how intricate is the analysis of job-search in UDCs' cities. The next section summarises some of the work carried out on the social, political and economic attitudes of those who are not successful in this search. In particular, questions like 'does prolonged and apparently irremediable unemployment inevitably lead to politicisation of labour' are considered.

Disillusion from Job-Search

P.C.W. Gutkind has spent many years documenting and interpreting the reactions of urban Africans who cannot secure jobs in prestigious, white-collar or regular occupations. Noting the common disillusion that 'no more is the standard six certificate a passport to that famed abode of tie and white shirt personalities; no more is the school certificate the golden ladder to that comfortable loft above, or the B.A. the key to the car garage',[54] Gutkind argues that the 'energy of despair' shades into the 'anger of despair', but that the transition is slow. And, 'what merit there is in these small-scale studies does seem to suggest that political consciousness increases, however modestly, with length of exposure to the conditions faced by the unemployed. Furthermore, the unemployed seem to shift their political focus from local conditions to a broader base — that of the nation as a whole'.[55]

During the 1950s and 1960s, many writers believed that a variety of class-consciousness was emerging amongst the urban poor in UDCs.[56] Others now believe that political consciousness among the new urban poor will not take the form of overt threats to the political status quo, however. Nelson distinguishes three possible reasons for this: first, the fact that most will have arrived from neighbouring towns, not the countryside, and, being acclimatised to urban life will not suffer such *anomie* or dislocation leading to anger; second, most will believe themselves to be better off in the city, and finally because those who are most discontented may leave by return-migration.[57] Naturally, interpreting such evidence as exists is difficult. It has been stated that in Latin America 'at least one source from every country surveyed stated that the squatters were more satisfied with their present housing and economic situation than with that they had had in the rural areas, small towns and in the central city'.[58] On the other hand, this may only be a temporary lack of political awareness: 'there has not yet generally emerged an urban-based status system that irrevocably anchors a worker in the matrix of new work situations'.[59] Nelson herself considers the possibility that the second-generation urban poor, who have not experienced their parents' dissatisfaction with smaller towns or the countryside, will be more emphatic in their criticisms.[60]

Clearly, assessing the nature and extent of political feelings amongst the urban poor is a hazardous task. First, one has simple problems of sampling: selecting only from those who are openly unemployed ignores the attitudes of those in petty trading and other occupations generally observed to have petit-bourgeois or some variety of 'self-help' ideology. Questioning itself introduces error to studies of this kind: people, particularly those without much experience in such matters, are likely to respond so as to please the interviewer — as they see it. Then the intensity of feeling cannot easily be gauged: one might reasonably feel a sympathy of some kind to a trade union, while simultaneously feeling rather more committed to a less overtly political institution such as a trade association. And, finally, difficulties of interpreting events in a political light hamper research. Do *ad hoc* formations of workers to oppose a government initiative or to secure a wage increase reflect politicisation in some sense, or are they purely instrumental gatherings of convenience which dissolve once the grievance is settled or no longer insupportable? For these reasons the study of incipient class consciousness and labour relations is still at a tentative stage. A selection from the writings confirms this.

Peace's study of Lagos observed that successful businessmen and women preferring to live in their accustomed neighbourhood had the effect of convincing others that they too could become wealthy.[61] The demonstration effect in such cities of rich and successful people confirming for others, so far less successful, that they too have a chance of success can be called 'con-mechs'. Elliott says of the successful:

They depend for their longrun survival on their ability to retain
the belief among competitors that thay have a chance – not
necessarily an 'even' or 'fair' chance, but an acceptable (or
acceptably fair) chance of winning the competition . . . The more
obviously and overtly competitive the mechanism, the more the
individual is likely to attribute failure to his own shortcomings.[62]

Assessing the question from a different angle, Elkan is sceptical
regarding the existence of a proletariat in Nairobi.[63] Citing as
evidence the continuing high rates of remittances to rural areas by even
long-established factory workers in Nairobi, he retains his earlier
position that for many East African workers urban life is only a phase,
to be ended after a number of years, and implying no strong commitment in political terms. In Nigeria, the 'cross-pressured worker' effect
is similar. Many workers there prefer to pay dues to an *association
d'originaires* than to a trade union, believing that the benefits they
stand to receive are far greater from the former than the latter.[64]
Evidence of mismanagement of trade union funds, a practice widely
suspected in Nigeria, further diminishes workers' willingness to join
unions.[65]

Yet a continuing allegiance to rural and ethnic ties on the part of
urban workers is not sufficient reason to dismiss the possibility that
urban and industrial-based class divisions may come to be widely
perceived. Sandbrook's review arguing that 'the notion of class
formation does not imply the absence of non-class identities' helps
retrieve the debate from an unilluminating dichotomy.[66] One
element of this topic which might particularly gain by recognising the
co-existence of different loci of interest is the class-consciousness of
the educated unemployed. Whether they wait patiently for a coveted
job in the 'formal' sector, exhibit radicalism for the duration of
their job-search and then adopt a more conservative outlook, or
whether they remain radical, is a subject much in need of research.[67]

Housing and Intra-Urban Movements

It is likely that recent migrants to cities in UDCs will move within
those cities several times. They may move because they wish to set up
their own house, or seek a better location to use for their roadside stall
or shop, or they may be looking for a lower rent to set up a
warehouse or factory. In all these cases, and also in response to
government low-cost housing initiatives, migrants, both recent and
long-established, will be moving inside towns and cities.

Flinn's study of Bogota discerned a typical two-stage move on the part of arrivals. Their first destination would be the centre of the city; their second the outskirts with their *tugurios* and *invasiones*.[68] Thus 28 per cent of migrants went straight to the city centre before moving out; and 58 per cent went to the fringe of the city. Those who went to the fringe first had rather more money with them.[69] In Peru, 87 per cent of a 1967 survey of shanty-dwellers did not use loans to buy their houses, and 13 per cent of those in Venezuelan *ranchos* owned their own shacks.[70] In addition, it is significant that only 8 per cent of the Bogota sample had neither friends nor relatives in the city they visited.[71]

An important reason for initially living in the city centre, close to those who may be hiring labour, is that job-search will have to be conducted frequently until some more regular subsistence is identified. Thus, sewage tips and rubbish dumps, if close to active parts of the city, will be used by 'footholders' or the newest arrivals. Those on the outskirts will, by contrast, have found more regular jobs. In Rio de Janeiro, there is a contrasting system of intra-urban movement, whereby the outer areas are visited first: 'The slum (inner city) is simply not a place where rural migrants come – probably because there is little space available, low turnover, much crowding and often prohibitions against children . . . Finally, although rates are cheap, there are still monthly payments to be made in contrast to the *favelas*, which are totally free.'[72]

By contrast, many West African migrants prefer only to rent accommodation. A recent hypothesis suggests that this behaviour is due to the customary practice of building a house in the town of origin before in the city; and because land can be acquired legally, obviating the need to squat without rights.[73] The desire to return home (one sample referred to 'our sons abroad') is explained by Peil in terms of access to family land at home and the possibilities for transferring small enterprises between towns once they are successfully underway. A similar explanation turned up in a study of Dahomean fishers living on Lagos lagoon: they spent 20 years or so in Nigeria (returning in December-February each year) before retiring to Dahomey permanently. Their land was rented from the nearby university: again, little capital was invested in their place of work.[74]

Indian cities do not exhibit a pattern of concentric rings, with slums in the centre, surrounded by the commercial area, then the peripheral shanties, as described by Andrews and Philips, (although

Delhi might be argued to fit this impression). In Calcutta and Eastern India the distinction is made between the 'basha' and the 'bustee'. The former ('home' in Bengali) is a house shared by friends while they are working in the city; the latter is slightly lower in status, often being on unauthorised sites. In view of the limited knowledge which exists regarding intra-urban movements in response to new housing and income conditions, Drakakis-Smith's conclusion is particularly apposite: 'more consideration has to be given to the microeconomics of the informal sector and the need to retain as many of the residential/economic ties as possible.'[75]

The following chapters will look at the micro-economics of the 'informal' sector: its genesis in development thought; the difficulties which arise from using it as a concept; the evolution of small enterprises; and growth and diversification prospects for the future.

Notes

1. R. Weitz (ed.), *Urbanization and the Developing Countries* (New York: Praeger, 1973), p. 13.
2. L. Wirth, 'Urbanism as a Way of Life', *American Journal of Sociology*, Vol. 44, July 1938, pp. 3-24.
3. T.G. McGee, *The Urbanization Process in the Third World* (London: G. Bell and Sons, 1971), pp. 25-52.
4. J. Abu-Lughod, quoted in Weitz, p. 48.
5. J. Aldous, 'Urbanization, the Extended Family and Kinship Ties in West Africa', in S.F. Favo (ed.), *Urbanism in World Perspective* (New York: Thomas Cromwell and Co., 1968), pp. 297-305.
6. J.E. Perlman, *The Myth of Marginality* (Berkeley: University of California Press, 1976).
7. S.O. Imoagene, 'Mechanisms of Immigrant Adjustment in a West African Urban Community', *Nigerian Journal of Economic and Social Studies*, Vol. 9, No. 1, March 1967, p. 53.
8. J. Gugler, 'Life in a Dual System: Eastern Nigerians in Town, 1961', *Cahiers d'Etudes Africaines*, Vol. 11, No. 3, 1971, pp. 400-21.
9. L. Jellinek, quoted in T.G. McGee, 'Hawkers and Hookers, Making Out in the Third World City. Some South-East Asian Examples', *Manpower and Unemployment Research*, Vol. 9, No. 1, April 1976, p. 9.
10. W. Mangin, 'Latin American Squatter Settlements: A Problem and a Solution', *Latin American Research Review*, Vol. 2, No. 3, Summer 1967, pp. 65-84.
11. P.L. Doughty, 'Behind the Back of the City: "Provincial Life" in Lima, Peru', in W. Mangin (ed.), *Peasants in Cities* (Boston: Houghton Mifflin, 1970), pp. 30-46.
12. Ibid., p. 32.
13. Loc. cit.
14. R. Albert Berry, 'Open Unemployment as a Social Problem in Urban

Colombia: Myth and Reality', *Economic Development and Cultural Change*, Vol. 23, No. 2, January 1975, p. 288.

15. Perlman, p. 80.
16. J. Ramos, 'An Heterodoxical Interpretation of the Employment Problem in Latin America', *World Development*, Vol. 2, No. 7, July 1974, p. 55, footnote 30.
17. Berry, p. 289.
18. Cited in Berry, p. 289.
19. Perlman, p. 80.
20. M.P. Todaro, *Internal Migration in Developing Countries* (Geneva: ILO, 1976), p. 14.
21. M. Peil, 'Unemployment in Banjul: The Farming/Tourist Tradeoff', *Manpower and Unemployment Research*, Vol. 10, No. 1, April 1977, pp. 25-9.
22. S.B.L. Nigam and H.W. Singer, 'Labour Turnover and Employment: Some Evidence from Kenya', *International Labour Review*, Vol. 110, No. 6, December 1974, pp. 479-93; and J.S. Henley and W.J. House, 'Collective Bargaining, Wage Determination and the Regulation of Employment Conditions in Kenya' (paper presented to Fourth World Congress of International Industrial Relations Association, Geneva, September 1976).
23. T.S. Papola and K.K. Subrahmanian, *Wage Structure and Labour Mobility in a Local Labour Market* (Bombay: Popular Prakashan, 1975), p. 126.
24. R.P. Dore, *The Diploma Disease* (London: George Allen and Unwin, 1976).
25. G.S. Fields, 'The Private Demand for Education in Relation to Labour Market Conditions in Less Developed Countries', *Economic Journal*, Vol. 84, December 1974, pp. 906-25.
26. P.B. Doeringer and M.J. Piore, *Internal Labour Markets and Manpower Analysis* (Lexington, Mass.: D.C. Heath, 1971).
27. W.F. Steel and Y. Takagi, 'The Intermediate Sector, Unemployment and the Employment-Output Conflict: A Multi-Sector Model' (Vanderbilt University Working Paper, Dept. of Economics, December 1976), p. 11.
28. D.Z. Zschock, 'Legal, Economic and Cultural Determinants of Labor Absorption by the Modern Sector: A Case Study of Manufacturing in Colombia', unpublished, 1976.
29. K. King, 'Kenya's Informal Machine Makers', *World Development*, Vol. 2, Nos. 4 and 5, April-May 1974, pp. 9-28.
30. K. King, 'Indo-African Skill Transfer in an East African Economy', *African Affairs*, Vol. 74, No. 296, January 1975, pp. 65-71.
31. Fields, p. 911.
32. C.C. Holt, 'Job Search, Phillips' Wage-Relation and Union Influence: Theory and Evidence', in E.S. Phelps (ed.), *Microeconomic Foundations of Employment and Inflation Theory* (London: Macmillan, 1970), pp. 53-123.
33. For example, Perlman, p. 62.
34. C. Elliott, *Patterns of Poverty in the Third World* (New York: Praeger, 1975), p. 305.
35. O.J. Fapohunda, 'Characteristics of the Unemployed People in Lagos' (Lagos: University of Lagos, Human Resources Research Unit, 1975), p. 18.
36. D.I. MacKay, 'After the "Shake-out" ', *Oxford Economic Papers*, Vol. 24, No. 1, March 1972, pp. 89-110.

37. J.T. Addison and J. Burton, 'Wage Adjustment Processes: A Synthetic Treatment', *Research in Labor Economics*, Vol. 1, No. 1, 1976.
38. E. Kritz and J. Ramos, 'The Measurement of Urban Underemployment', *International Labour Review*, Vol. 113, No. 1, January-February 1976, pp. 115-27.
39. H. Joshi and V. Joshi, *Surplus Labour and the City: A Study of Bombay* (London: Oxford University Press, 1976), p. 172.
40. Ibid., p. 134.
41. Perlman, p. 81.
42. Papola and Subrahmanian, p.111.
43. Berry, p. 287.
44. Fapohunda, p. 27.
45. K. King, 'Kenya's Educated Unemployed: A Preliminary Report', *Manpower and Unemployment Research in Africa*, Vol. 7, No. 2, November 1974, pp. 45-63.
46. T. Boyd and S. French, 'A Summary of Research Findings on Secondary School-leaver Unemployment in Ghana, 1969–1971', *Manpower and Unemployment Research in Africa*, Vol. 6, No. 1, April 1973, pp. 56-61.
47. S.N. Sen, cited in B. Dasgupta, 'Calcutta's Informal Sector', *Institute of Development Studies Bulletin*, Vol. 5, No. 2, October 1973, p. 55.
48. Joshi and Joshi, p. 133.
49. IBRD, 'Some Aspects of Unskilled Labour Markets for Civil Construction in India: Observations based on Field Investigations' (Washington DC: World Bank Staff Working Paper No. 223, November 1975), p. 6.
50. Papola and Subrahmanian, pp. 111-14.
51. M.A. Bienefeld and E.M. Godfrey, 'Measuring Unemployment and the Informal Sector: Some Conceptual and Statistical Problems', *Institute of Development Studies Bulletin*, Vol. 7, No. 3, October 1975, p. 9.
52. C. Gerry, 'The Wrong Side of the Factory Gate: Casual Workers and Capitalist Industry in Dakar, Senegal', *Manpower and Unemployment Research*, Vol. 9, No. 2, November 1976, p. 22.
53. Perlman, p. 7.
54. P.C.W. Gutkind, 'From the Energy of Despair to the Anger of Despair: The Transition from Social Circulation to Political Consciousness among the Urban Poor in Africa', *Canadian Journal of African Studies*, Vol. 7, No. 2, 1973, p. 180.
55. Ibid., p. 196.
56. S. Amin, 'The Class Struggle in Africa', *Revolution*, Vol. 1, No. 9, 1964; (published under the pseudonym of XXX): K.W. Grundy, 'The Class Struggle in Africa: An Examination of Conflicting Theories', *Journal of Modern African Studies*, Vol. 2, No. 3, November 1964, pp. 379-94.
57. J.M. Nelson, *Migrants, Urban Poverty and Instability in Developing Nations* (Cambridge, Mass.: Center for International Affairs, Harvard University, Occasional Paper No. 22, 1969).
58. Mangin, p. 89.
59. P.C.W. Gutkind, *The Poor in Urban Africa: A Prologue to Modernization, Conflict and the Unfinished Revolution* (Montreal: Centre for Developing Area Studies, McGill University, 1968), p. 165.
60. Nelson, p. 36.
61. A. Peace, 'The Lagos Proletariat: Labour Aristocrats or Populist Militants', in R. Sandbrook and R. Cohen (eds.), *The Development of an African Working Class* (London: Longman, 1975), pp. 291-302.
62. Elliott, p. 11.
63. W. Elkan, 'Is a Proletariat Emerging in Nairobi?', *Economic Development*

and Cultural Change, Vol. 24, No. 4, July 1976, pp. 695-706.

64. R. Melson, 'Ideology and Consistency: The "Cross-Pressured" Nigerian Worker', in R. Melson and H. Wolpe (eds.), *Nigeria: Modernization and the Politics of Communalism* (Michigan: Michigan State University Press, 1971), pp. 581-605.

65. M.O. Kayode, 'The Management of Trade Union Finances in Nigeria', in U.G. Damachi and H.D. Seibel (eds.), *Social Change and Economic Development in Nigeria* (New York: Praeger, 1973), pp. 138-46.

66. R. Sandbrook, 'The Study of the African "Sub-Proletariat": A Review Article', *Manpower and Unemployment Research,* Vol. 10, No. 1, April 1977, pp. 91-105.

67. Ibid., p. 99.

68. W.L. Flinn, 'The Process of Migration to a Shanty Town in Bogota, Colombia', *Journal of Inter-American Economic Affairs,* Vol. 22, No. 2, Autumn 1968, pp. 77-88.

69. F.M. Andrews and G.W. Philips, 'The Squatters of Lima: Who they are and what they want', *Journal of Developing Areas,* Vol. 4, No. 2, January 1970, pp. 211-24.

70. C. Abrams, 'Squatting in Venezuela', in A.R. Desai and S.D. Pillai (eds.), *Slums and Urbanization* (Bombay: Popular Prakashan, 1970), p. 108.

71. Flinn, p. 86.

72. Perlman, p. 19.

73. M. Peil, 'African Squatter Settlements: A Comparative Study', *Urban Studies,* Vol. 13, 1976, pp. 155-66.

74. S.W. Sinclair, 'Ease of Entry into Small-Scale Trading in African Cities: Some Case Studies from Lagos', *Manpower and Unemployment Research,* Vol. 10, No. 1, April 1977, pp. 79-90.

75. D.W. Drakakis-Smith, 'Urban Renewal in an Asian Context: A Case Study in Hong Kong', *Urban Studies,* Vol. 13, No. 3, October 1976, p. 304.

4 THE URBAN EMPLOYMENT QUESTION

The previous chapter showed how, on arrival in towns and cities, migrants begin to look for work. In certain respects they behave similarly to those born therein. Both groups are required to find subsistence, but the speed with which they initiate this search, the intensity with which they conduct it, and the eventual outcome, will differ between age groups and ethnic groups, and by level of education and other factors. Chapter 2 pointed out that migration is selective: those who move to towns and cities are not a representative sample of the people who live in the places they leave. Migrants tend to be young and well-educated relative to those they leave behind — although the markedness of this varies between countries. They also tend to retain ties of various kinds with those left behind; these can be species of 'support networks', cash relations, or claims to building and farming land. Another important feature of these migrants is that by no means all of them will stay for the rest of their lives in the place they move to. Those who move on voluntarily may return to their place of origin after a predetermined time has elapsed or sum accumulated; they may alternatively move on to a still larger city. Those who leave involuntarily will not have found sufficient support and will have consequently abandoned their planned stay in the city.

This chapter is concerned with the failure of the manufacturing and service sectors of the 'modern' economy in UDCs' cities to absorb the labour supplies to which they have had access. The consequence of this failure — 'the employment crisis' — has been the subject of a huge literature. This will be summarised before chapter 5 examines the growth of the so-called 'informal' sector, which is one of the forms in which the urban unemployed have responded to this employment crisis. It will become apparent in this chapter that it is misleading to talk of 'an employment problem'. Quite apart from the difficulties residing in the term 'employment' itself (these were discussed in chapter 1), it is more illuminating to think of not one employment problem, but many. A number of factors give rise to slow rates of labour absorption in UDCs' cities, and while policy action can be expected to mitigate some of these trends, it cannot help others. And those people who are affected respond in many different ways. Thus, it is important that the heterogeneity of peoples' experiences and cities' developments is recognised. Emphasis on the 'urban poor' as if they were an identifiable and homogenous group is bound to hamper analysis.

The 'Formal' Sector in the Economy

The 'formal' sector in UDCs' cities is a term used with varying degrees of precision to refer to large-scale manufacturing plant, industrial capacity which uses identifiably 'Western' technology, office work and services which are similarly organised along 'Western' lines. Characteristics usually ascribed to this sector or group of enterprises include the use of accounting methods, the payment of regular wages and salaries, the use of a fixed place of business, and some fixed capital. None of these characteristics unambiguously differentiates 'formal' business from the residual 'informal' sector: even the smallest and most unsophisticated street-trader might exhibit some of the characteristics listed, albeit in indigenous form.[1] (The ambiguity of the distinction between these two 'sectors' is the concern of chapters 5 and 6.) But notwithstanding the definitional and conceptual difficulties of the terms 'formal' or 'modern', they are useful at a very simple and aggregated level of description. As a starting-point for understanding why these enterprises have failed to use as much labour as has been available in cities, Table 4.1 shows the extent of their demand for labour. It can be seen from column 3 that, in most UDCs, the 'formal' sector accounts for only a small proportion of the labour force. Thus, for this small island to expand to absorb the entire increment to the labour force each year would require a staggering increase in employment by each plant. If all plants together employ 10 per cent of the labour force in a city, and that labour force is growing, due to migration and natural internal increase, at 5 per cent annually, 50 per cent annual rises in 'formal' employment are needed. (As chapter 1 pointed out, the notion of a 'labour force' is rather unclear. The labour force can be defined in terms of those who are willing to work at any wage above subsistence, or those who are willing to work only at certain 'reservation' wage levels, and so on. Here the term is used with reference to a rough estimate of those employed and unemployed. The statistics collected in most UDCs are not comprehensive enough to permit of a more rigorous definition.) This projected increase in demand for labour would have to be even greater if account were taken of increases in productivity. As workers become more used to a job ('the learning curve'), their output per unit of labour-time rises, meaning that increases in output require

Table 4.1: The Growth of Manufacturing Employment and the Labour Force

Region/Countries	Employees in manufacturing (annual growth rate 1963-9)	GNP per capita (annual growth rate 1960-70)	Manufacturing labour force as % of total labour force 1970
East Africa			
Botswana	–	–	1.0
Ethiopia	6.4	2.8	–
Kenya	–	3.6	–
Malawi	7.1[b]	2.1	–
Mauritius	1.4[d]	–	14.6
Somalia	–18.0[d]	–1.1	–
Sudan	–	1.0	5.0
Uganda	–	2.4	8.0
United Republic of Tanzània	14.0[e]	3.6	–
Zaire	–	2.7	3.1
Zambia	15.3[f]	7.1	2.6
West Africa			
Gabon	–	–	1.9
Ghana	6.3	–0.4	8.6
Ivory Coast	–	4.5	0.8
Liberia	–	0.9	2.1
Niger	–	–2.0	0.1
Nigeria	5.7[f]	0.1	–
Sierra Leone	–	4.7	4.4
Asia			
China, Republic of	13.3[c]	7.1	–
Fiji Islands	–	–	7.0[a]
Hong Kong	9.2	–	41.4
India	–	1.2	9.5
Indonesia	–	1.0	5.6
Khmer, Republic of	–	0.1	2.7
Korea, Republic of	13.0	6.8	13.2
Malaysia (East Sabah)	12.3[d]	–	–
Malaysia (West)	8.1[c]	3.1	8.7
Nepal	–	0.5	1.9
Pakistan	2.6[f]	2.4[h]	9.5
Philippines	4.8	2.9	11.4
Singapore	17.4	–	13.9
Sri Lanka	–	1.5	9.1
Thailand	–12.0[g]	4.9	3.4
Latin America and the Caribbean			
Argentina	–	2.5	25.1
Barbados	–	–	14.6
Bolivia	–17.0[e]	2.5	10.3

Table 4.1 cont.

Table 4.1: *cont.*

Region/Countries	Employees in manufacturing (annual growth rate 1963-9)	GNP per capita (annual growth rate 1960-70)	Manufacturing labour force as % of total labour force 1970
Brazil	1.1	2.4	17.8[b]
British Honduras	—	—	14.1
Chile	4.2[g]	1.6	23.2
Colombia	2.8[f]	1.7	12.8
Costa Rica	2.8[f]	3.2	11.5
Dominican Republic	−3.3[f]	0.5	8.2
Ecuador	6.0	1.7	14.0
El Salvador	—	1.7	12.8
Guadaloupe	—	—	10.4
Guatemala	—	2.0	11.4
Guyana	—	—	15.1
Haiti	—	−0.9	4.9
Honduras	10.6[f]	1.8	7.8
Jamaica	—	—	13.7
Martinique	—	—	8.8
Mexico	—	3.7	16.7[b]
Netherlands Antilles	—	—	25.8
Nicaragua	—	2.8	12.0
Panama	7.4	4.2	7.6
Paraguay	—	1.3	15.1
Peru	—	1.4	13.2
Puerto Rico	—	—	17.2
Surinam	—	—	8.9
Trinidad and Tobago	20.0[c]	1.9	14.7
Uruguay	—	−0.4	21.6
Venezuela	—	2.3	18.6

a 1966
b 1970
c 1966-9
d 1967-9
e 1966-8
f 1963-8
g 1963-7
h Includes Bangladesh

Source: D. Morawetz, pp. 492-5.

proportionately smaller increases in labour input. Counteracting this is the recorded tendency for labour-force participation rates (LFPR) to fall with urbanisation. LFPRs measure the proportion of those willing to work to the population as a whole. Males under 20 years of age and over 65, in particular, exhibit a declining LFPR as they move to the city; while females' behaviour is less consistent and varies by region. Broadly, female LFPRs are higher for non-agricultural than agricultural areas in Latin America, but higher for agricultural areas in South and East Asia. Another broad trend is the males' LFPRs to be lower, the higher the per capita income of the country. For females, however, a consistent pattern is not discernible.[2]

Given the rate of urban migration and population increase recorded, it is not altogether surprising that varieties of open unemployment have emerged. However, the point remains that even the island of industrialisation which has appeared might have used more labour per unit of output had different technologies been selected and installed; had different products been made; and had some different policies regarding the wages-structure, the exchange-rate and subsidies to capital been implemented.[3]

Technology and Employment

The choice of technology depends on a number of factors, of which the most important are: information on the availability of alternatives; the relative prices of different types of labour, and of different types of installation and imported components – both on credit and for cash; the availability of supervisory labour to oversee unskilled labour; the product to be made; whether or not the enterprise is foreign-owned; and tastes.

To take these *seriatim*: it is not clear, first of all, how much knowledge is available regarding technology options. Since only about 1 per cent of world research and development (R&D) into productive techniques is carried out in UDCs and even some of that is related to problems of most interest to rich countries ('the internal brain drain'), there is little indigenous invention. But there may be adaptation, along the lines used by Japanese entrepreneurs after the Meiji restoration.[4] In this process, second-hand or discarded equipment is used to produce with less than 'latest-generation' techniques, but with techniques well-suited to local factor endowments. By embracing second-hand and superseded machinery from DCs, enterprises in UDCs are able to extend the 'shelf' of technologies to which they have access.[5]

The importance of relative factor prices on the choice of techniques is much disputed. Based on the experience of South East Asian UDCs, Ranis's work shows considerably sensitivity of new plants to factor-price-ratios.[6] (The issue of changing the technology used by extant plants is rather different. Concentration on this to the exclusion of looking to the future sometimes makes the discussion of technology choice overly gloomy.) A related point is that enhanced productivity (output/labour-input) of labour after acculturation to a job can make labour more attractive than capital after a while, even at apparently disadvantageous initial relative prices. Policy adopted by UDCs' governments can affect the factor-price ratio. Subsidies to capital effected through an overvalued exchange rate, allowing capital goods and components to be imported relatively cheaply, is a common characteristic of policy in UDCs.

The impact of public policy on wage-rates (especially the institution of minimum wage legislation) and trade union rights is also a contentious subject. The importance of rising real wages for the substitution of capital for labour depends critically upon the elasticity of substitution. If this is high, instituting minimum wage legislation will cause demand for labour in extant plants to drop, as substitution takes place; or demand for labour in plants built in the future to be lower than in existing plants. Testing in UDCs to estimate the elasticity of substitution is difficult and the results obtained often ambiguous. The assumption of homogeneity of labour and capital results in 'all-or-nothing' analysis in which recognition of shifts toward more use of supervisory labour, or greater use of on-the-job training, cannot be accommodated. Substitutability also depends upon how far the product-mix can be changed without jeopardising total revenue, and upon leeway in the intensity of capital-use. Many plants in UDCs operate only one shift per day: willingness to make more use of the idle capital implied by this excess capacity will depend upon the type of entrepreneur involved and the supply of co-operant inputs — notably supervisory labour.[7] There is, finally, the difficulty of extricating causal movements through time. Higher wages paid in respect of greater output per unit of labour-input may be a result of greater capital investment rather than simply a cause.

Estimates of the elasticity of substitution of capital for labour rarely exceed unity. There are case studies with elasticities in excess of 1 — grain-milling, paint and woollen textiles production in India[8] — and others finding elasticities beneath 1.[9] The results obtained will depend upon the size of plant tested, and are, furthermore, jeopardised by use of constant elasticity of substitution production functions, which involve limiting assumptions such as perfectly competitive markets, and constant returns to scale.

The importance of wage increases in changing the factor proportions of a plant will depend on many variables. The effectiveness of minimum wage legislation and other statutory wage measures naturally plays a role. A survey of wage legislation by Watanabe found that government measures of that sort were effective in very few UDCs.[10] The conclusions of Reynolds and Gregory's well-known study of Puerto Rico to the effect that employers respond quickly and decisively to changes in wages by diminishing their employment levels, were found not to be easily generalisable.[11] If the impact of such legislation is felt disproportionately by large firms, whose small number, ease of identification and (perhaps) political sensitivity make them prime targets for wage legislation, there will tend to be differential elasticities of substitution by size of plant. Another factor which might lead to greater capital-intensity of operation in large plants is the likelihood of their having an in-house or plant union, rather than a nationally-organised union, negotiating. To the extent that such unions have an interest in maximising company profits so as to maximise wages and bonuses, there may be a stimulus to substitution of capital for labour.[12]

The whole subject is bedevilled by interpretative difficulties. A major difficulty lies in attributing causality between observations through time. The well-known conundrum of whether unions in negotiation can ever be furnaces (causing wage increases) rather than merely thermometers (registering and rubber-stamping wage increases) is amongst the problems present.

A more useful insight into labour-capital substitution is possible when 'labour' is split into unskilled and skilled or supervisory components. The relative cost and abundance of each will affect the capital-intensity of operation. A firm which has expensive capital installations will obviously be unwilling to entrust these to unskilled labour alone: some overseers are needed. But if these overseers are unobtainable, the price of the unskilled labour becomes irrelevant. The importance of labour costs in a sophisticated installation would have been slight anyway, but in such a case the attractions of unskilled labour are eroded by the absence of supervisors. Related to examples such as these is the possibility that too low a wage differential between unskilled and skilled manual labour, to the extent that this results in a shortage of skilled labour before their market wage has time to respond upwards, results in less unskilled labour being used. Thus, rather than labour being too expensive, in crude terms, it is conceivable that too low a reward for a certain type of labour para-doxically reduces employment.

Whether or not a plant is foreign-owned can be important in determining the technology used. Multi-national corporations are accused of using standard technologies wherever they operate, irrespective of local factor-price ratios or resource availabilities. But country-level studies have identified exceptions. A comparative study of American and indigenous firms in Mexico and the Philippines found that American firms were only marginally more capital-intensive, after plant size and product composition were accounted for.[13] Agarwal's study of Indian manufacturing discovered greater capital-labour ratios, in 22 out of 34 factories, being used by foreign-owned firms, but found the difference in capital-intensity to be small, and possibly explicable by the typically larger size of plant used by foreign firms.[14] One factor allowing foreign firms to use more capital was their easier access to the capital market. Small and as yet uncreditworthy domestic entrepreneurs were unlikely to be able to borrow at advantageous rates of interest in UDCs' towns and cities with fragmented capital markets.

A different type of impact deriving from foreign ownership is knowledge and awareness of available technologies. Managers in large firms (especially multi-national corporations) may be particularly informed in this respect, and may therefore be more innovative.

An important reorientation of the foreign/domestic ownership dichotomy for the analysis of capital-intensity of production has introduced the REF (resident expatriate-owned firm).[15] Data from Ghanaian manufacturing plants indicated that 'the important bloc of manufacturing units owned by resident expatriates shows much greater affinity with the multinationals than with the truly indigenous firms.' Thus, empirical work should in future distinguish these firms from both wholly-foreign and wholly-indigenously owned firms for the purposes of examining factor use. One might extend this tripartite approach to include a further variety of manufacturing unit, which is increasingly common in UDCs: the joint venture, owned in part by the foreign operator, the host government and indigenous shareholders. These too are likely to have different predilections for capital and labour.

Just as different methods of combining labour and capital to make an object can be devised, so can the nature of the object be changed to use more of one factor of production. To make a given product, there is some scope for substitution between factors — as the previous section discussed. But there is usually greater scope for such substitution in the ancillary or peripheral operations, such as warehousing, packaging and receiving. Such differences in the way a product is produced may not, as in the case of receiving and warehousing, affect the nature of the product as it is perceived by the buyer; but packaging does affect its nature.[16] The extent to which production may so be altered must depend on the nature of the demand. If demand is linked closely to the distribution of income, then changing the product may severely curtail its marketability. If, for instance, the high-income echelons of a community require their loaves to be presented in sealed plastic bags and their shirts to be packaged in smart cardboard and vinyl containers, the possibilities for labour-intensive handling are constrained. On the other hand, if items are still attractive when packaged by hand, then clearly there is scope for greater labour-use.

A particular variety of output with important employment implications is manufactured exports. Although they are concentrated in the higher-income UDCs, with the exception of India (that is, in Brazil, Hong Kong, South Korea, Taiwan), these exports have been growing very rapidly in the last ten years. Between 1972 and 1973 the manufactured exports of the most succcessful eleven UDCs increased by 54 per cent in value terms.[17] But these exports need not be more capital-intensive than domestic production. Mexico's exports of labour-intensive craft goods are growing just as fast as her exports of capital-intensive goods.[18] South Korea's exports from small-scale units (employing ten or fewer people) rose from 18.6 per cent of total manufactured exports in 1963 to 31.4 per cent in 1968.[19] Taiwan's industrial employment grew at 8.1 per cent annually during the 1960s. Conditions of access for manufactured goods will be renegotiated in the 'Tokyo Round' of General Agreement on Tariffs and Trade (GATT) negotiations, and may eventually present a very promising outlet for labour-using manufactures.

A final factor which impinges upon the nature of the technology installed in LDCs is taste. Prestige is felt in certain firms or industries to reside in the latest or most expensive technology, and foreign advisers have been accused of exacerbating this tendency. The latest technology for a given process developed in rich countries tends to be labour-saving and thus the phenomenon of taste transfer buttresses the problems already referred to in this chapter.[20]

This brief review of the conditions giving rise to the lack of sufficient demand for labour by 'formal' firms in cities adds to the picture developed in chapter 3. In that chapter the nature of hiring preferences and job-seekers' attitudes were discussed; in chapters 5 and 6 the alternative sources of subsistence used by unsuccessful job-seekers in the 'formal' sectors are examined.

Notes

1. S.W. Sinclair, 'Informal Economic Activity in African Cities', *Journal of Modern African Studies*, Vol. 14, No. 4, December 1976, pp. 696-9.
2. J. Durand, *The Labor Force in Economic Development* (Princeton: Princeton University Press, 1975), pp. 150-60.
3. A comprehensive survey is provided in D. Morawetz, 'Employment Implications of Industrialization in Developing Countries', *Economic Journal*, No. 335, Vol. 84, September 1974, pp. 491-542.
4. S. Paine, 'Lessons for LDCs from Japan's Experience with Labour Commitment and Subcontracting in the Manufacturing Sector', *Bulletin of Oxford University Institute of Economics and Statistics*, Vol. 33, No. 2, May 1971, pp. 115-34.
5. A.S. Bhalla, 'Small Industry, Technology Transfer and Labour Absorption', in OECD, *Transfer of Technology for Small Industries* (Paris: OECD, 1974), pp. 107-20.
6. G. Ranis, 'Some Observations on the Economic Framework for Optimum LDC Utilisation of Technology', in L.J. White (ed.), *Technology, Employment and Development* (Princeton: Princeton University Press, 1974), pp. 58-96.
7. G.C. Winston, 'The Theory of Capital Utilisation and Idleness', *Journal of Economic Literature*, Vol. 12, No. 4, December 1975, pp. 1301-20.
8. H. Pack, 'The Employment-Output Trade-off in LDCs — A Microeconomic Approach', *Oxford Economic Papers*, Vol. 26, No. 3, November 1974, pp. 388-404.
9. R.L. Williams, 'Capital-Labor Substitution in Jamaican Manufacturing', *The Developing Economies*, Vol. 12, No. 2, June 1974, pp. 169-81.
10. S. Watanabe, 'Minimum Wages in Developing Countries: Myth and Reality', *International Labour Review*, Vol. 113, No. 3, May-June 1976, pp. 345-58.
11. L.G. Reynolds and P. Gregory, *Wages, Productivity and Industrialization in Puerto Rico* (Homewood, Illinois: Irwin, 1965).
12. R. Dore, 'The Labour Market and Patterns of Employment in the Wage Sector of LDCs: Implications for the Volume of Employment Generated', *World Development*, Vol. 2, Nos. 4 and 5, April-May, 1974, pp. 1-7.
13. R. Hal Mason, 'The Transfer of Technology and the Factor Proportions Problem: The Philippines and Mexico' (New York: UNITAR Research Report, No. 10, 1971).
14. J.P. Agarwal, 'Factor Proportions in Foreign and Domestic Firms in Indian Manufacturing', *Economic Journal*, Vol. 86, No. 343, September 1976, pp. 589-94.
15. D.J.C. Forsyth and R.F. Solomon, 'Choice of Technology and Nationality of Ownership in Manufacturing in a Developing Country', *Oxford Economic Papers*, Vol. 29, No. 2, July 1977, pp. 258-82.
16. H. Pack, 'The Substitution of Labour for Capital in Kenyan Manufacturing', *Economic Journal*, Vol. 86, March 1976, pp. 45-58.
17. UNCTAD, *Trade in Manufactures of Developing Countries and Territories, 1974 Review* (New York: UNCTAD, 1976), p. 8.
18. R.W. Boatler, 'Trade Theory Predictions and the Growth of Mexico's Manufactured Exports', *Economic Development and Cultural Change*, Vol. 23, No. 8, April 1975, pp. 491-506.
19. G. Ranis, 'Industrial Sector Labor Absorption', *Economic Development and Cultural Change*, Vol. 21, No. 3, April 1973, pp. 387-408.
20. O. Hawrylyshyn, 'Non-Economic Biases towards Capital-Intensive Techniques in LDCs' (Queens University, Ontario, Dept. of Economics Discussion Paper No. 233, n.d.).

5 THE EMERGENCE OF THE 'INFORMAL' SECTOR

The previous chapter showed how, as a response to lack of work in 'formal' occupations, and, indeed, because of certain preferences too, people in UDCs' cities turn to find subsistence in small-scale enterprises run by themselves, friends, relatives or strangers. The myriad forms this response has taken have emerged as one blanket term in the literature – the 'informal' sector. This chapter will begin by tracing the genesis of the term, and will then examine its various guises in different countries. The theme of the chapter is that the 'informal' sector as a concept in development studies requires considerable refinement, both conceptually and in the light of empirical evidence. This leads to chapter 6, in which proposals for different ways of seeing the 'informal' sector are discussed.

The Background to the 'Informal' Sector

Before tracing the emergence of the 'informal' sector concept in development studies, it may be useful to begin with a description of how at least one writer has interpreted it:

> . . . the roadside and empty-lot mechanics who will weld on a Bourneville Cocoa tin to mend the exhaust pipe of the civil servant's Mercedes, the leather workers making hand-made bags for the tourist trade, the furniture-makers, the men who collect empty Essolube cans from the garages twice a day and have them processed into serviceable oil-lamps by sunset.[1]

The foregoing account cannot be thought of as definitive; different writers would stress other aspects of the 'informal' sector: perhaps the tendency to smallness of size of plant or shop, its physical dispersion throughout a city, its relationships with larger firms or the state. But niceties of definition and analysis will be discussed at greater length later.

The progenitor of the two-sector labour transfer model in UDCs was the celebrated Lewis model of 1954.[2] This popularised the concept of dualism, in which a small industrial sector grows to absorb greater amounts of rural labour, which can be lost to cities without any corresponding drop in agricultural output. By paying a constant real wage at subsistence level, and which is less than the marginal product of labour, industrialists are able to re-invest a surplus and employ an increasing proportion of the UDC's labour force. Movement from the rural sector takes place at a constant real wage (although Lewis raised the possibility of a 30 per cent or so money wage differential to account for higher city prices) and in response to certain employment. There is no urban unemployment in this model, merely rural under-employment or hidden unemployment; nor is there any mention of an 'informal' sector.

As elaboration of one aspect of Lewis's model was provided by Ranis and Fei,[3] who presented a more detailed account of consumption in the two sectors. Other modifications have also appeared: Weeks has reinterpreted the Lewis model in terms of labour exploitation;[4] Frank has attacked it on the grounds of its containing an overly fragmented view of UDCs;[5] and it has been criticised for containing of necessity a 'hidden' agricultural revolution to ensure a rising marketable food surplus for sale to urban workers.[6] But of all these amendments, the theme of central importance to this chapter is the appearance of a succession of intermediate sectors between the city and the countryside.

Apart from Gutkind's anthropological studies of the dispossessed unemployed in African cities,[7] the first model embracing more than two sectors — the rural (which was assumed for simplicity to be wholly agricultural) and the urban (which was characterised by industry) — was Reynolds's.[8] In this model he explicitly introduced two further urban groups: the state sector and the 'trade-service' sector. The latter he described as 'the multitude of people whom one sees thronging the city streets, sidewalks and back alleys in the LDCs: the petty traders, street vendors, coolies and porters, small artisans, messengers, barbers, shoe-shine boys and personal servants'.[9] This 'trade-service' sector drew upon Geertz's earlier work on the 'bazaar' economy. In this short analysis of two Indonesian towns, Geertz indicated how market trade was a general mode of commercial activity — 'the expression of [the trader's] essential self'.[10] The anthropological observations there were analysed by Reynolds to demonstrate why and how the number of market traders and casual labourers might grow as a response to such factors as migration, the speed of labour-absorption in industry, and so on. Immediately after Reynolds's work a spate of similar papers appeared. Most of them were exercises in typologising — classifying sectors of the labour market in more or less different ways. It was not until the emergence of the 'informal' sector

debate that the focus of attention shifted from classification to
rigorous analysis. Some of the earlier work is reviewed first.

Scoville's typology of Afghanistan urban labour markets[11] included
three sectors: skilled and unskilled 'traditional' sectors, and an
'internal labour market' or industrial sector. This was primarily an
attempt to show that traditional incomes need not necessarily be
beneath modern incomes: if skills were highly developed, they could
readily earn higher incomes than were found in the lower echelons of
the industrial sector. Frankman and Charle's analysis of sub-Saharan
Africa adopted the term 'service sector' to apply to traders, service
purveyors and the like. This sector was seen as a 'training ground for
increasingly satisfying occupations . . . and more responsible
positions'.[12] Hinchliffe's study of Northern Nigeria examined the
idea that there existed an 'aristocracy of labour' in industry, enjoying
substantial and secure advantages of income and prestige over those
outside the industrial sector.[13] He found that no such advantages
were enjoyed, and that to a large degree the higher money wages of
industrial workers were absorbed in higher prices obtaining near their
places of work. Henley and House, in their study of Kenyan labourers,
also adopted a strict aristocracy of wage-labour and 'informal' sector
dichotomy, on the basis of wage-rate and labour turnover data.[14]
They hypothesised that low rates of voluntary turnover in factories
indicated satisfaction with conditions therein, but this phenomenon
could, of course, also be interpreted as fear of not finding a comparable alternative job in the face of substantial unemployment of
school-leavers. John Weeks's typology rested upon the 'rich' sector
and the 'poor' sector: 'the rich sector is an extension, one might even
say an invasion, by the rich countries into the poor countries. This
sector is alien to the society and economy; its capital, skills and
techniques are imported.'[15] Here again there is a straightforward
dichotomy between the advanced-technology sector employing a small
and highly privileged proportion of the industrial labour force, and
the residual, which in this case is called the 'poor' sector. So far in the
literature, then, all the people who could not be counted as regular
wage-employees in plants of sufficient size to be enumerated in
industrial censuses were treated as one mass. Very few characteristics
were ascribed to them, other than their exclusion (by definition)
from secure, and (by implication) respectable employment.

Hart's seminal study of what he called the 'informal' sector in
the township of Nima in Ghana presented for the first time the
heterogeneity of this group of people who had previously been

lumped together as a residual from agriculture and industry.[16] He set out with the criticism that terms like 'underemployed' or 'traditional', when used to describe the urban poor, beggar analysis by assuming what has to be demonstrated. A useful explanatory and predictive analysis must begin by fully recognising the many different occupations which these so-called 'informal' people perform, how regularly and to what end they perform them, and the myriad ways in which economic and social life is blended in shanty towns.

Shortly afterwards, the ILO report on Kenya, which embraced the 'informal sector' as a major part of its analysis, defined it thus: 'informal activities are a way of doing things, characterised by (a) ease of entry; (b) reliance on indigenous resources; (c) family ownership of enterprises; (d) small scale of operation; (e) labour-intensive and adopted technology; (f) skills acquired outside the formal school system; (g) unregulated and competitive markets.'[17] The usefulness of this collection of enterprises, not only in absorbing and supporting urban workers, but in fostering entrepreneurship, lowering the costs of living by providing simple goods at low profit margins for other urban workers (this was a point quickly seized by Marxist critics of the report) and reducing the need to import manufactured goods, was noted. This represented a complete reversal of attitudes towards these purveyors. A report in 1960 referred to: 'the problem presented by the large number of Africans who are seemingly unemployed [and] itinerant hucksters who take advantage . . . of the large quantities of produce being brought in by gullible peasants to act as middlemen, and so bedevil the marketing of produce.'[18] In fact Nairobi City Council as late as 1972 still strongly opposed stall-holders and other operators' failure to acquire licences or use official sites. The clash in that year with outdoor barbers showed the extent of the Council's intransigence, as did the destruction in 1970 of 7,000 'unauthorised' houses.[19]

The reorientation of development thought brought about by the appearance of the 'informal' sector was desirable in two main ways. First it indicated the contribution which those in its various parts made to other parts of the economy. The suggestions that entrepreneurship was fostered therein, that imports were saved, that skills were acquired and that rural households gained from the remittances of part of urban 'informal' workers' incomes all led to a change in the way these people were perceived. Just as a decade before squatter townships began to be seen as 'solutions' rather than 'problems', so the unorganised part of the city economy came to be appreciated rather than reviled. But apart from this somewhat functionalist perspective, the appearance of an 'informal' sector economics demonstrated the importance of greater attention being paid to the ways in which people behaved and earned a living in this fastest growing part of UDCs' cities. The relatively unrefined first treatment of the mass of pedlars and manufacturers in the late 1960s and early 1970s stimulated a rich new vein of criticism and exegesis, and it is to these criticisms that the chapter now turns.

Doubts about the 'Informal' Sector

The criticisms of the 'informal' sector can be analysed under three headings: aggregation, linkages and dynamics.

Among the first to criticise the concept of the 'informal' sector on the grounds that it is too highly aggregated to permit useful analysis to take place were Bienefeld and Godfrey, who wrote:

> . . . the so-called 'informal sector' includes a large variety of people and activities situated very differently in relation to each of these issues [market power and degree of domination by larger units] : it is therefore essential that the sector under discussion be substantially disaggregated in such a way that its components become analytically significant and that each can be defined in a way which is statistically useful.[20]

As a solution to the difficulty of overly aggregated analysis, they proposed a tripartite division into 'those activities which produce tradeable commodities, activities which involve the production of services connected with distribution and finance, and activities which involve personal services'.[21] Yet even this tripartite division obscures certain important phenomena. For instance, the processes whereby some small enterprises fail and are usurped by others, or whereby a petty retailer grows to become a wholesaler, are not illuminated by a disaggregation along the lines suggested there. Elsewhere, Bienefeld has written of the need to split up the sector on the basis of occupations if progress is to be made in identifying those parts of the city economy where initial absorption and assimilation of new entrants to the labour-market takes place.[22] His survey of Dar-es-Salaam tentatively concluded that only street-hawking performed the role of absorbing recent arrivals to the city, and that in general the 'sector' did not act as employer of last resort for all who turned to it. Elkan too has argued that aggregating the multitude of urban activities into 'an informal sector is not only a piece of needless obscurantism but also raises the question whether these activities do in fact constitute a sector'.[23]

A more recent ILO World Development Programme survey than that performed in Kenya — that for the Sudan in 1976 — argued similarly that 'our understanding of the informal sector would be enhanced if we viewed it as a heterogeneous, multidimensional or multilayered phenonemon'.[24] To this end they distinguished four sub-groups in the category, which are then analysed quantitatively. They conclude:

> The picture that emerges for the informal sector in the Sudan is that of heterogeneous and complex activities. At its most advanced level we have the well-established enterprises carrying out the bulk of retail trade. At its middle level, where the majority of establishments exist, we have the multitude of small manufacturing, service and commercial establishments employing a large number of people who are making a reasonable living and who are there to stay. Finally we have the traditional petty vendors who are in transition to and from formal-sector jobs and who at the moment do not seem to constitute a significant portion of the Sudanese informal sector.[25]

It seems a little incongruous that such a careful analysis should conclude with the term 'informal sector' in the singular: one would have expected 'sectors' in the plural, given what is quoted before. But the thrust of the report's conclusions is clear — that splitting up the 'sector' is necessary before progress can be made in understanding its components and the different forces which each component faces. Similarly, Breman's critique argues:

> . . . instead of applying the concepts of formal-informal, we should distinguish in terms of different articulated production relations which can be found within the economic system of third world countries.[26]

The proposed resolution takes the form of 'graduations' rather than a strict dichotomy: 'the structure of this [labour] market is not dualistic, but has a far more complex ranking.'[27]

Thus the first difficulty encountered in the use of the 'informal' sector concerns the homogeneity it apparently ascribes to all the enterprises and individuals lumped within it. The problems of this residual approach are that it becomes hard to identify which parts of the 'informal' sector have growth potential, which parts are irremediably doomed to ekeing out a subsistence standard of living, which parts are readily penetrated by newcomers and which parts only with difficulty, and finally the differences in prospects and style of operations between families, firms and individuals therein. A more illuminating mode of analysis will therefore be less embracing in its scope, and will begin with a more micro-economic approach. The precise ways in which this can proceed are discussed in chapter 6; this chapter continues with the other criticisms which might be directed at the 'informal' sector.

The second criticism concerns the linkages between the 'informal' part of the urban economy and the rest. One of the conclusions of the ILO Kenya report was that closer relations between the 'formal' and 'informal' portions of the economy should be fostered.[28] It was envisaged that, by a variety of subcontracting and other relationships, patterns of demand hitherto restricted to formal suppliers could be harnessed to pull along the small enterprises, and enable them to escape the low demand/low savings/low investment/low productivity nexus in which they were imagined to be languishing. Thus, it was suggested that car assembly plants should make an effort to buy some of their components (for example, windscreen wipers and seat frames) from small indigenous firms, rather than importing them all, or ordering them exclusively from offshoots of developed country component manufacturers. In this way small firms and the labour they used would benefit from the enclave type of development so often criticised for worsening the distribution of income in UDCs as growth of income per capita begins to rise. The enclave of a small group of workers making expensive goods for a small group of high-income consumers would thus be broadened to the advantage of all.

Yet two aspects of such linkages were not fully thought out by the report. First is Leys's concern that the subcontracting relationships envisaged would lead eventually to the demise of the smaller partner, for increasing indebtedness and the need to purchase inputs and sell outputs to better-organised and better-financed firms would lead 'informal' firms into being exploited and to possible bankruptcy.[29] Leys's fears have been substantiated by surveys carried out among Calcutta shoemakers. In a survey of 640 small enterprises, Bose concluded:

. . . the qualitative picture which emerges from these case studies
fully confirms the hypothesis that the dominating large-scale
oligopolistic sector compels the small units in the informal sector
to operate in a different market where the input price is higher and
the output price is lower, and the main benefit of this price
differential is reaped by the large-scale sector.[30]

He cited instances such as sandal-retailers who bought sandals from
small-scale manufacturers at 11 rupees, slightly less than the cost of
production, packaged them attractively but at low unit cost, and sold
them at 23 rupees. Calcutta has some 2,000 sandal-making units,
altogether providing subsistence for some 10,000 people. It is a
seasonal business, with the number of units involved swelling to 3,000
or so just before festivals, when demand for sandals rises.[31] The
industry illustrates well the pitfalls involved in a policy which seeks to
expand earnings in the 'informal' portion of an occupation. Large-scale
suppliers of inputs ('gaddiwalas'), three in number, are able to control
the leather and adhesive supplies to the small units. And, after
fabrication, retailers such as Bata Shoes buy in bulk at advantageous
prices. A wholesaler will buy on behalf of retailers, sometimes
managing to drive down the price obtained by the fabricator so far
that no profit whatsoever is obtained. What enables small makers to
remain solvent and continue providing their participants with subsistence
is the occasional sale of fancy items at higher prices and at much higher
unit profits. Recognising that large dealers would be able to
requisition small manufacturers' entire capacity for months on end,
and would therefore be able to drive very hard bargains, the ILO's
World Employment Programme Guidelines conceded that 'encourage-
ment of subcontracting must live with this implication'.[32] Thus closer
links cannot be assumed to be good *per se*. In another context, Leys
has emphasised that analyses which ignore the distribution of market
power between two negotiators are likely to be inadequate: 'so many
of the determinants of income distribution are left out (if this is done)
that one must feel grave doubts that the policies proposed would, even
if pursued, in themselves prove sufficient to make significant impact'.[33]
    A further study of linkages between large and 'informal' parts of
the economy of Dakar underlines these difficulties. Gerry's study
covered 285 enterprises in four occupations: shoemakers, carpenters,
tailors and vehicle-repairmen.[34] He found that 79 per cent of the small
firms in furniture-making (subsumed under carpenters) were wholly
reliant, for their raw material inputs, upon oligopolistic suppliers.

While a tiny number of small firms received lucrative state contracts (often by cultivating patronage), the 'dominant position' of larger-scale capital could not be circumscribed. He concluded, along with Leys and Bose, that the strategy of strengthening linkages between 'informal' firms and others was unlikely to result in substantially enhanced incomes for many of the former.

To achieve a more thorough understanding of the inter-relationships which exist in the city much more care has therefore to be devoted to specifying the ways in which 'informal' units reply upon and mingle with other units. It is clearly not sufficient to assume that all parts of the city economy can co-exist amicably and indefinitely. Each part of the economy will be threatened, circumscribed or enhanced by its interactions with other parts – this requires investigation. The fact that the 'informal' sector concept in the form in which the ILO Kenya report cast it does not throw any light upon the effect of linkages reflects the third shortcoming of the concept. This is its lack of a dynamic which can provide an understanding of how these enterprises are supposed to grow or fail. In part this is a reflection of the degree of aggregation of the concept, as outlined above. For, naturally, with such a heterogeneous assemblage of operators, there can be no single production function and no single growth pattern which may be predicted. In fact, this shortcoming has led one commentator to reject the concept outright: 'the informal sector approach . . . warrants no further discussion'.[35] He further argued:

> . . . a framework in which the productive ensemble is viewed as an ensemble . . . would tend to focus on the relationships between the different elements of the ensemble, rather than stress the mutually exclusive characteristics of one element vis-à-vis another . . . One quite crucial result arises from the utilisation of an approach which tends not to use an initial *a priori* dichotomy . . . [i.e.] constraints facing the petty producer and trader will be seen to be derived not from the individual characteristics of the 'enterprise' itself . . . but rather the global structure of urban production and distribution.[36]

An attempt to devise a method for analysing and predicting the future growth of various 'informal' occupations by Mosley uses the ILO Kenya report as a source of data.[37] The analysis proceeds on the basis of estimating the income-elasticity of demand for the output of each occupation, and then projecting their likely prospects under various rates of growth of income. It might be argued that this method is somewhat flawed, since it assumes that people in one occupation are unable to move between occupations or product lines as demand patterns change. This is often an unjustifiable assumption, as chapter 2 has shown. Nevertheless, Mosley's work represents an important beginning in work trying to compensate for the lack of dynamics in the 'informal' sector literature in its simpler formulations. Leys has also attempted this, asking how far patterns of demand might already have been forged by rich sector tastes, 'so that any substantial increase in the incomes of the working poor would expand demand for "formal sector" products (such as bread, leather shoes, bicycles, etc.) rather than for "craft-produced" cheap goods'.[38] Some empirical evidence on this last point is provided by Langdon's study of the Kenyan soap industry.[39] Langdon found that the growth and eventually the survival of indigenous, small-scale soap producers was threatened by the entry of Unilever's marketed brands which appealed to low-income consumers largely on the basis of attractive packaging.

Still another approach to studying the dynamics of 'informal' enterprise is that of Harriss, whose South Indian evidence cuts across the 'formal'/'informal' distinction completely.[40] She found many instances of activities which would usually be considered 'informal' on an intuitive basis (for example, illegality, the use of relatives and friends rather than tendering in the market to take and give orders) being carried on at the same time as holding 'formal' occupations. Thus, people might be keen to secure 'formal' sector jobs because of their subsequent ability simultaneously to operate 'informal' enterprises in their spare time, or to cultivate contacts for use in later self-employment. This type of behaviour she calls 'quasi-formal', by way of suggesting that typologies of economic behaviour are no substitute for disaggregated dynamic analysis. In similar vein, Moser has studied preferences for different occupations, and has found that these preferences change through an individual's life.[41] Again, this suggests that the two types of economic behaviour cannot be usefully dichotomised. An interesting insight into job preference is Bujra's study of prostitution in early Nairobi, which documents women's willingness to enter that occupation as a prerequisite to accumulating substantial capital and eventually becoming major property owners.[42] Here the prostitute, explicitly cited by Hart as an 'informal' sector operator, and firmly included in the urban 'lumpenproletariat' by Marx, Fanon, Gutkind and others, becomes instead the social and economic climber par excellence. Finally, the growth patterns of small enterprises is argued by Allen to depend upon access to capital: this, in fact, is his criterion for determining the boundaries of the 'informal' sector.[43] Because of their using illegal land, and their inability to save due to keen competition keeping prices low, Kenya's car-repairmen have severely limited growth prospects, he argues.

The foregoing paragraphs have outlined three major shortcomings of the 'informal' sector concept. First, by the treatment of its constituents as a 'sector', with the attendant implied shared production function, analysis of each individual part of the whole is impeded. Which parts of the 'sector' may easily be entered is not clear; which parts have hopeful growth prospects, and when, and in what countries, is similarly unclear. Second, the relationships these operators have with others in the city economy is left unexplored; yet whether these relationships are useful 'linkages' in Hirschman's sense,[44] or exploitative in others' senses is essential to decide if policies are to be formulated for this sector. Third, the dynamics of these operators through time is not made clear by the conventional treatment of the 'informal' sector. The way in which its elements may grow or change, in connection with which other elements of the urban milieu, is not illuminated.

These, then, are the three major difficulties presented by the 'informal' sector concept. While recognition of this group of urban workers is surely an advance over the previous concern with under-employment and hidden unemployment, it should not be accepted uncritically. Chapter 6 will present alternatives to the approaches which have just been criticised. Before finishing, however, the topic of entrepreneurship in the 'informal' sector has to be mentioned. In these small enterprises it is natural that administrative, managerial and marketing skills should play an important role. Yet entrepreneurship is a nebulous concept, even in received theory in developed countries. What entrepreneurs do (maximise or satisfice or both, subject to other constraints), how they do it, and why, are subjects of a considerable literature in the theory of the firm. The following brief review of the application of this work to the 'informal' sector illuminates these difficulties.

Entrepreneurship

The contributions of McClelland and Hagen have been among the most influential. McClelland's measure of 'n-Achievement' purports to show differences in willingness and ability between individuals to 'achieve', although, as critics have pointed out, precisely *what* is to be achieved is left somewhat unclear.[45] Hagen's 'withdrawal of status respect' hypothesis has been influential in the attempt to explain why alien minority groups are often more successful than their indigenous peers: one thinks of East African Asians, or Levantines in West Africa.[46] The central hypothesis here is that 'a derogated group will gradually turn from the values of its derogators . . . if the derogators are traditional, and economic prowess is a promising channel of escape . . . then the derogated may turn to economic activity'.[47] An empirical application of the latter theory is that of Levine,[48] who claims to have found that 'dream achievement' in Nigeria suggests Ibos and Southern Yoruba are the most interested in business success. The Hausa are less so because they believe their culture is — and is perceived to be — relatively unviolated by Western penetration. Winder's study of the Lebanese in Ghana and Nigeria confirms their desire for success,[49] but studies of this sort tend to be virtually irrefutable, claiming that certain groups' success can be explained in terms of their possession of prerequisites for success. Kennedy's study of Ghana attempts to delineate those factors which compel people into 'petty bourgeois independence'.[50] These he takes to be the preference for self-employment in West Africa, and the constant turn-over of the contracted industrial labour-force, which gives rise to a pool of unemployed labour, made redundant by technological progress. This preference for self-employment has also been documented by Lloyd.[51] Beveridge and Oberschall's study of Lusaka businessmen concludes that the business 'climate' (a rather nebulous term in itself) may be crucial in determining an individual entrepreneur's success.[52] By way of indirectly confirming Kennedy's finding that small scale of operation is frequent, they ascertain that nine out of the fifteen industrialists they surveyed found troubles with employees the most vexing they faced. They therefore preferred a small labour force. Marris and Somerset used 1966-7 data in their study of Kenyan businesses.[53] They identified a desire for independence: 'the vocational dissatisfaction which drives men into business derives its idealism from a prevailing sense of frustrated national aspirations. Entrepreneur-ship is a reaction to the economic backwardness which excludes African society as a whole from international respect.'[54]

This interpretation apparently rests on an implicit 'withdrawal of status respect' view (they mention Kikuyu business drive as consistent with their analysis, the Kikuyu having been somewhat reviled in the colonial years). More has been said about preferences for different sorts

of jobs in chapters 3 and 4. As for smallness of firm, Marris and
Somerset cite reasons including employers' desire to avoid alleged theft
and laziness by employees, and the preference on the part of
successful businessmen for buying land rather than diversifying into
other commercial interests in town.[55] A study of land-use in Nigeria
by Peil makes a similar point – that wealth is more often used to
acquire real-estate in the town or village of origin, not in the city in
which the wealth is generated.[56] Since ownership rights to village land
typically weaken as the length of absence increases, many town-
dwellers in Uganda, for instance, plan to return home to retire.[57] The
small size of indigenous firms in Africa has also been noted by Weeks
and Liedholm,[58] and partially explained in terms of risk-aversion, or
fear at taking on too many overheads (supervisory labour to oversee
unskilled workers, bigger factory or workshop, more circulating capital,
and so on). A World Bank study of African entrepreneurship concluded
similarly that there was a strong aversion to expanding a firm beyond
one-man operations. Reluctance to delegate authority was seen as the
reason. In Kenya, this was found to give rise to multiple companies
being run by one man: all but 10 of a sample of 44 businessmen there
operated more than one firm.[59] Finally, a sample of 269 Nigerian
firms employing ten or more people was unable to resolve the problem
of why so many enterprises remain small.[60] Preference for intense
individualism and imperfect capital markets, which constrain
borrowing, were suggested as partial explanations. The use of 'extended
families', in which patterns of obligation and privilege affect all
members, was also proffered as a cause; although, paradoxically, this
same social network has been found, in other studies, to make possible
the starting of new businesses.[61] Nafziger's survey found that about
three-quarters of his sample of 28 firms had used personal and/or
family savings to get the business launched.

The conclusions of this chapter follow from its analysis of the
evolution of the 'informal' sector concept, and its account of some of
the criticisms which have been levelled against it. If recognition is to be
given in development studies to the mass of urban small-scale traders,
manufacturers and service purveyors, the vehicle of analysis must be
chosen carefully. First, studies must eschew universality, and attempt
to explain only parts of this mass at a time. *A priori* generalisations
regarding the growth of 'informal' firms, ease of entry to them, and
their relations with others in the UDC, are likely to be at best misleading, and at worst plain wrong. Disaggregation of various sorts is
therefore a prerequisite to an understanding of this important part of

the urban economy. Next, exploration of these operators' relationships with larger and similar operators must proceed. Only when an understanding of the 'informal' sector's relationships is attained can policies regarding employment, income distribution, migration, incomes policy, and so on, be adequately formulated. And third, the ways in which 'informal' firms can be expected to change through time must be researched. Static ascription to categories is not useful for this task; it only serves to emphasise the enormous problems of classification. What is needed, then, is not a Linnaeus of the 'informal' sector's statics, so much as a Schumpeter of its dynamics.

Notes

1. R.P. Dore, *The Diploma Disease* (London: Unwin, 1976), p. 74.
2. W.A. Lewis, 'Economic Development with Unlimited Supplies of Labour', *Manchester School*, Vol. 22, No. 2, May 1954, pp. 139-91.
3. G. Ranis and J.C.H. Fei, *Development of the Labor Surplus Economy* (Homewood. Illinois: Irwin, 1964).
4. J. Weeks, 'The Political Economy of Labour Transfer', *Science and Society*, 1971, pp. 463-70.
5. A.G. Frank, *Sociology of Development and Under-development of Sociology* (London: Pluto Press, 1971), esp. pp. 41-3.
6. G. Myrdal, *Asian Drama* (Harmondsworth: Penguin, 1968), Appendix 6.
7. P.C.W. Gutkind, 'African Responses to Urban Wage Employment', *International Labour Review*, Vol. 97, No. 2, February 1968, pp. 135-66; *The Poor in Urban Africa: A Prologue to Modernization, Conflict and the Unfinished Revolution* (Montreal: Centre for Developing Area Studies, 1968); later works include 'From the Energy of Despair to the Anger of Despair: The Transition from Social Circulation to Political Consciousness among the Urban Poor in Africa', *Canadian Journal of African Studies*, Vol. 7, No. 2, 1973, pp. 174-98; *The Emergent African Urban Proletariat* (Montreal: Occasional Paper No. 8, Centre for Developing Area Studies, 1974).
8. L.G. Reynolds, 'Economic Development with Surplus Labour: Some Complications', *Oxford Economic Papers*, Vol. 21, No. 1, March 1969, pp. 89-103.
9. Ibid., p. 91.
10. C. Geertz, *Pedlars and Princes* (Chicago: University of Chicago Press, 1963), p. 44.
11. J.G. Scoville, 'Afghan Labour Markets: A Model of Interdependence', *Industrial Relations*, Vol. 13, No. 3, October 1974, pp. 274-87.
12. M. Frankman and E. Charle, 'Employment in the Service Sector in Sub-Saharan Africa', *Journal of Modern African Studies*, Vol. 11, No. 2, 1973, pp. 201-10.
13. K. Hinchcliffe, 'Labour Aristocracy: A Northern Nigerian Case Study', *Journal of Modern African Studies*, Vol. 12, No. 1, 1974, pp. 57-67.
14. J.S. Henley and W.J. House, 'Collective Bargaining, Wage Determination and the Regulation of Employment Conditions in Kenya' (paper presented

to Fourth World Congress of the International Industrial Relations Association, Geneva, 6-10 September 1976).

15. J. Weeks, 'An Exploration into the Nature of the Problem of Urban Imbalance in Africa', *Manpower and Unemployment Research in Africa*, Vol. 6, No. 2, November 1977, pp. 9-36. The quotation is from p. 20.

16. K. Hart, 'Informal Income Opportunities and the Structure of Urban Employment in Ghana', *Journal of Modern African Studies*, Vol. 11, No. 1, March 1973, pp. 61-89.

17. ILO, *Employment, Income and Equality: A Strategy for Increasing Productive Employment in Kenya* (Geneva: ILO, 1972), pp. 23-6.

18. Colonial report, quoted in P. Mosley, 'Implicit Models and Policy Recommendations: Reflections on the Employment Problem in Kenya' (paper presented to British Institute of Geographers, London, March 1977).

19. H.H. Werlin, 'The Informal Sector: The Implications of the ILO's Study of Kenya', *African Studies Review*, Vol. 17, No. 1, April 1974, pp. 205-12.

20. M.A. Bienefeld and M. Godfrey, 'Measuring Unemployment and the Informal Sector', *Institute of Development Studies Bulletin*, Vol. 7, No. 3, October 1975, p. 8.

21. Ibid., p. 8.

22. M.A. Bienefeld, 'The Self-Employed of Urban Tanzania' (Institute of Development Studies Working Paper, No. 17, n.d.), p. 25.

23. W. Elkan, 'Concepts in the Description of African Economies', *Journal of Modern African Studies*, Vol. 14, No. 4, December 1976, pp. 691-5. Quotation is on p. 693.

24. ILO, *Growth, Employment and Equity: A Comprehensive Strategy for the Sudan* (ILO: Geneva, 1976), p. 315.

25. Ibid., p. 386.

26. J. Breman, 'A Dualistic Labour System? A Critique of the "Informal Sector" Concept', *Economic and Political Weekly* (Bombay), 27 November, 4 December, 11 December 1976 (3 parts), p. 1875.

27. Ibid., p. 1905.

28. ILO, pp. 5, 225, 259.

29. C. Leys, 'Interpreting African Underdevelopment: Reflections on the ILO Report on Employment, Incomes and Equality in Kenya', *Manpower and Unemployment Research in Africa*, Vol. 7, No. 2, November 1974, pp. 19-28, esp. p. 25.

30. A.N. Bose, 'The Informal Sector in the Calcutta Metropolitan Economy' (Geneva: ILO Working Paper, 1974).

31. T. Basu, 'Calcutta's Sandal Makers', *Economic and Political Weekly* (Bombay), 6 August 1977, p. 1262.

32. ILO, *World Employment Programme: Research in Retrospect and Prospect* (Geneva: ILO, 1976), p. 160.

33. C. Leys, 'The Politics of Redistribution with Growth', *Institute of Development Studies Bulletin*, Vol. 7, No. 2, 1975, pp. 4-8.

34. C. Gerry, 'Petty Producers and the Urban Economy: A Case Study of Dakar' (Geneva: ILO Working Paper, No. 8, 1974).

35. C. Gerry, 'Petty Production and Capitalist Production in Dakar: The Crisis of the Self-Employed' (paper presented to British Institute of Geographers, London, 19 March 1977), p. 1.

36. Gerry (1974), pp. 6-9, 109-110.

37. Mosley, pp. 9-11.

38. C. Leys (1974), p. 28, footnote 17.

39. S. Langdon, 'Multinational Corporations, Taste Transfer and Under-
    development: A Case Study from Kenya', *Review of African Political
    Economy*, No. 2, 1975, pp. 12-35.
40. B. Harriss, 'Quasi-formal Employment Structures and Behaviour in the
    Unorganised Urban Economy, and the Reverse: Some Evidence from
    South India' (paper presented to British Institute of Geographers,
    London, 19 March 1977).
41. C. Moser, unpublished paper.
42. J.M. Bujra, 'Women "Entrepreneurs" of Early Nairobi', *Canadian Journal
    of African Studies*, Vol. 9, No. 2, 1975, pp. 213-34.
43. H. Allen, 'The Urban Informal/Industrial Sector and Growth: Some
    Thoughts on a Modern Mythology' (paper presented at 12th Annual
    Social Science Conference of East African Universities, 1976, Dar-es-
    Salaam).
44. A.O. Hirschman, *The Strategy of Economic Development* (New Haven:
    Yale University Press, 1958).
45. D.C. McClelland, *The Achieving Society* (Princeton, N.J.: Van Nostrand
    and Co., 1961).
46. E.E. Hagen, *On the Theory of Social Change* (Homewood, Illinois:
    Dorsey Press, 1962).
47. E.E. Hagen, *The Economics of Development* (Homewood, Illinois: Irwin,
    1975), p. 284, footnote 11.
48. R.A. Levine, *Dreams and Deeds: Achievement Motivation in Nigeria*
    (Chicago: Chicago University Press, 1966).
49. R. Bayley Winder, 'The Lebanese in West Africa', *Comparative Studies in
    Society and History*, Vol. 4, No. 3, April 1962, pp. 296-333.
50. P. Kennedy, 'Cultural Factors Affecting Entrepreneurship and
    Development in the Informal Economy in Ghana', *Institute of Develop-
    ment Studies Bulletin*, Vol. 8, No. 2, September 1976, pp. 17-21.
51. P.C. Lloyd, *Power and Independence* (London: Routledge and Kegan
    Paul, 1974), p. 154.
52. A.A. Beveridge and A. Oberschall, 'African Businessmen in Lusaka: Some
    Initial Findings' (paper presented at East African Universities Social
    Science Conference, Dar-es-Salaam, December 1970).
53. P. Marris and A. Somerset, *African Businessmen: A Study of Entre-
    preneurship and Development in Kenya* (London: Routledge and Kegan
    Paul, 1971).
54. Ibid., pp. 63-9. Quotation is from p. 69.
55. Ibid., pp. 109-12, 118.
56. M. Peil, 'African Squatter Settlements: A Comparative Study', *Urban
    Studies*, Vol. 13, No. 2, June 1976, pp. 155-66.
57. J. Odongo and J.P. Lea, 'Home Ownership and Rural-Urban Links in
    Uganda', *Journal of Modern African Studies,* Vol. 15, No. 1, March 1977,
    pp. 59-73.
58. J. Weeks, unpublished MS; and C. Liedholm, 'Research on Employment
    in the Rural Non-Farm Sector in Africa' (African Rural Employment
    Study, Michigan State University, African Rural Employment Paper
    No. 5, 1973), pp. 8-9.
59. IBRD, 'The Development of African Private Enterprise' (Washington DC:
    World Bank Report AW.31, 1971), pp. 15, 122-6.
60. J.R. Harris, 'On the Concept of Entrepreneurship, with an Application to
    Nigeria', in S.P. Schatz (ed.), *South of the Sahara: Development in African
    Economies* (London: Macmillan, 1972), pp. 5-27.
61. E. Wayne Nafziger, 'The Effect of the Nigerian Extended Family on

Entrepreneurial Activity', *Economic Development and Cultural Change*, Vol. 18, No. 1, October 1969, pp. 25-33.

6 SHORTCOMINGS OF THE 'INFORMAL' SECTOR

From the criticisms presented in chapter 5, a number of hypotheses about the 'informal' sector can be derived. The argument contained in that chapter, stated briefly, was that the use of one concept — *the* 'informal' sector — impedes analysis. First, it implies a greater degree of homogeneity regarding those within that sector than is actually justified in the light of empirical studies; second, the term obscures and does not allow examination of the linkages which exist between different parts of the 'informal' sector's components and other parts of the city's economy; and thirdly, it does not facilitate analysis of the dynamic forces at work in changing the 'informal' sector's parts through time. This chapter carries on the analysis of the sector by examining constraints on entry to the occupations implied by the term, and growth prospects for those inside it. Entry can be over-conceptualised, just as 'employment' can. If a task is undertaken for a few hours each week, or different tasks are undertaken in successive weeks, one might talk of 'entry' into casual labouring, but this would be unnecessarily cumbersome. But for traders, some types of self-employed service purveyors and manufacturers, there are definable constraints on the initiation of business, and these are analysed below.

Ease of Entry

The belief that entry into informal occupations is easy is widespread. Thus,

> an important characteristic of this unit is its very low capital (fixed and working) requirements. The retailer usually obtains his goods on credit from the expatriate firm. He does not carry a stock since this is usually replenished at short intervals . . . entry into distributive trade is easy because of low capital outlay.[1]

little capital is required to get started. Second, it is possible to be viable and competitive at small size. Third, there are no technological advantages. Fourth, since food retailing is local in nature, personal factors are important. Finally, the level of know-how required is within easy reach of illiterate people.[2]

In other continents, similar assertions have been made: for Djakarta, 'entry into such activities is practically unrestricted';[3] Reynolds's general survey spoke of its being 'a natural entry point',[4] and Merrick's survey of Brazil concluded: 'the informal sector clearly plays the buffer role ascribed to it by segmentation models.'[5] (By 'segmentation models' he refers to those models of UDCs' labour markets which explicitly introduce heterogeneous labour supply and conditions of hiring; the 'buffer' being the unprotected or unorganised portion of the labour market.)

It is important to know whether or not these assertions are true, particularly for understanding migration. If subsistence can easily be acquired, one may expect migration to cities to be heavier than if entry into 'informal' occupations is hazardous. The analysis of income distribution in UDCs is assisted by some idea of the ease with which cash may be gained in these occupations, too (although 'income' is an unwieldy concept itself, as are 'wages' and 'employment', in economies with substantial use of barter, non-cash payment for labour services, and irregular, part-time or fitful patterns of work). This was the conclusion of Rao's contribution to the *Redistribution with Growth* study.[6] The two biggest limitations he identified in research-ing the urban poor were, inevitably, the definitional and conceptual problems of 'poverty'; and the need for recognising poverty as a household rather than an individual phenomenon. Secondary workers in each household will make substantial differences to the gross earnings of the unit. A prerequisite to such research is careful analysis of earnings and entry to 'informal' occupations.

The present author surveyed 100 market traders in Lagos in early 1976, in an attempt to assess how easily they had been able to begin work. In accordance with the principles set out in chapter 5, the sample was split up by type of goods sold, to enable differences in conditions to be perceived.[7] It was found that few market-traders began to trade as soon as they arrived in Lagos. (Only 12 out of 100 were born in Lagos.) Fish-sellers experienced an average delay of five years, shirt-sellers three years, and street-hawkers two years. The reasons for this delay differ between products. In the case of fish, the sellers were faced with some institutional obligations — the need to join the market association, pay some fees, find a good site, and so on. At the other extreme, street-hawkers, who use no fixed stall and require only enough working capital to last a day or so, experienced no barriers to beginning work. The tendency for them to be younger and less well-educated (in terms of years of formal schooling) was also marked.

A variety of other sources confirm the view that entry is not straightforward. For West African trading, Adekokunna has noted that a fee is usually payable when entering a market.[8] The regulation of markets is often rigorous, despite their occasionally anarchic appearance. Little has documented the need to win the favour of the market queen,[9] and occasionally the need to contribute to a communal rotating credit fund (or *esusu* in Yoruba).[10] Similarly, market women in Abidjan, Ivory Coast, are rigorously controlled, being obliged to join associations with frequent meetings and powers of censure if unsatisfactory trade behaviour is found.[11] Vagale's study of Ibadan markets stresses the degree of regulation there.[12] Originally, markets in Yorubaland were owned by the chiefs in front of whose residences the markets were established; the majority are still owned by individuals or families. Even in the markets where traders pay no rent, they are not permitted to erect shops or stalls at will.

From other continents, there is further supporting evidence of the difficulty of entering 'informal' occupations. In Djakarta, a residence permit must be acquired before street-trading can be practised;[13] similar restrictions apply to Cali's 13,000 street-traders. And Latin American markets are also regulated:

> Many Latin American markets are notable for their orderly arrangement of traders and stalls in rows. Municipal police or market tax collectors often circulate in the marketplace and attempt to persuade or coerce traders into straight rows or designated areas . . . agreements between traders . . . are often reinforced by the authority and arbitration of formal or informal traders' associations.[14]

Breman's study of South Gujarat, India, established that the 'fragmented' labour market was a more useful method of arranging evidence than the 'informal'/'formal' dichotomy.[15] But even with this schema he found that the lowest 'port of entry' was not amenable to immediate entry. There was found to be much less mobility between street occupations than is often believed: shoeshiners, for instance, in Patna, must rent space from an intermediary who has in turn leased the rights from the municipal authorities. A box must be rented if an apprenticeship is sought — the box contains cleaning materials — and this entails paying half of one's takings to the master. Bonds of this sort continue for years if the apprentice is unably to pay off his debt fully.

Pakistan's street-cleaners were analysed by Streefland, and were found to face similar constraints on beginning work and subsequently on consolidating their position (promotion or diversification).[16] Streefland divided the entry constraints into four sets. First are those deriving from control of resources. If one did not own the requisite means of production (tools for a mechanic, boxes for a shoeshine man, stalls for street-sellers), one had either to borrow money or begin as an apprentice of some kind. Second was the 'urban environment', by which he meant changes in labour market conditions. A move to expel street-hawkers from public view, for instance, would alter the demand conditions facing the occupation. Next was government policy, the importance of which in assisting or harrying small operators has already been mentioned. Finally was 'capitalist penetration', meaning the products and patterns of labour use which accompany the extension of capitalist patterns of manufacture and distribution into a traditionally less marketing-conscious society. This useful four-part division of entry constraints does tend to obscure conflicts of interest within occupations, but is nonetheless a contribution to the under- standing of small enterprise.

Some of Streefland's points are well illustrated by Newcombe's study of food hawkers in Hong Kong.[17] These were increasingly squeezed out of their accustomed selling sites during the fieldwork described in his article. The rapidly growing number of supermarkets, which were less convenient locationally and more expensive for poor consumers, favoured the interests of higher-income residents. Here 'capitalist penetration', assisted by government interference, had the effect of severely curtailing business prospects for stall holders.

Conditions of Entry for New Migrants

An important distinction can be drawn between conditions of entry to various occupations for long-established town residents as against conditions facing new arrivals. One might expect there to be differences of this sort for three reasons. First the new arrival is likely to have less sophisticated knowledge of the labour market and fewer contacts. (Although, it will be recalled from chapter 3, this is not always the case. Often people have arranged their jobs in the town before leaving their previous place of residence.) Second, to the extent that there is a discernible hierarchy of occupations within the 'informal' sector in a town, the new arrival is more likely to be offered jobs or niches in the lower echelons, left available as the residents have worked their way into better positions. This implies, of course, a structure of opportunity which rewards persistence in the labour market with improved pay and conditions, and this view may turn out to be empirically false upon examination. But it must at least be tested. Third, if migrants are from racially identifiable areas, discrimination of various types may be applied against them. If this acts to limit the jobs they can take and their prospects of increased earnings within those jobs, their conditions of entry will once again differ from those facing urban residents.

Three recent studies throw some light upon this question. A survey of migrants to San Salvador City and towns in Guyana established that those who moved from rural rather than urban areas to a given city enjoyed less occupational mobility within that final city of residence.[18] Thirty per cent of those from rural areas who had previously had steady sources of subsistence were unable to find any work at all after moving, whereas this was true for only 13 per cent of the urban to urban migrants. The latter were, in addition, able to leave the least desirable urban jobs more quickly than the rest, and graduate to higher-paid work. They concluded that the hypothesis that migrant status significantly affected chances of a 'formal' job is true. Contrasting evidence comes from Sao Paulo, Brazil. There it was found that after six years or so migrants' earnings had risen to be approximately equal to those of non-migrants. This was despite the significantly smaller proportion of migrants with educational certification.[19] And in a Zambian shanty town, Van Velsen found that most of the inhabitants who had moved there in recent years were employed in regular work. Only 14 per cent were defined as self-employed, and only 7 per cent were categorised as 'unemployed'.[20] Not enough work has yet been carried out on this subject, but there is enough to show that migrants to towns do not necessarily find it impossible to acquire jobs and earnings comparable with those held by urban residents.

The 'Informal' Sector as Employer of Last Resort

Having looked at the question of ease of entry, it is now necessary to assess the desirability of that entry; in other words, are occupations outside the 'informal' sector always preferred to those within it? And, if so, why? Most of the literature on urban labour markets ascribes to 'informal' occupations the role of absorber of all the people who were unable to find work elsewhere in the city. How far this is justified is assessed in the following section.

The standard models of job-search, evolved by human capital

theorists and others attempting to explore the micro-economics of
unemployment theory in developed countries, are based upon the
assumption of declining aspirations through time. These aspirations
may be related to income, occupation, prestige, or some combination
of these, and the decline may be linear through time, or otherwise.
When modified for use in the cities of under-developed countries,
however, there has been a tendency in this framework to
(implicitly or otherwise) set 'informal' activities at the bottom of
the aspirations spectrum — that is, to see these activities as only
considered as a last resort, and as being less desirable than
virtually any other source of subsistence. To some extent, there has
been good empirical reason for doing so. Studies of the relationship
between migration, education and job-search have shown, not
surprisingly, that in the short term, those with more educational
credentials are less keen to take spare-time, part-time or other
'informal' work than are those with fewer credentials (see chapter 3).
And, in addition, it is well known that jobs in 'formal' enterprises
are growing less rapidly than the urban labour-force (see chapter 4).
For simple survival, many people are therefore obliged to seek
'informal' work. In addition to those who have never found or sought
'formal' jobs, these numbers will be swelled by those who once held
such jobs but lost them: victims of labour-saving technological advance,
or of bankruptcies of their own or others' firms. This group
approximates to Marx's proletariat:

> the lower strata of the middle-class — the small tradespeople,
> shopkeepers and retired tradesmen generally, the handicraftsmen
> and peasants — all these sink gradually into the proletariat,
> partly because their diminutive capital does not suffice for the
> scale on which moden industry is carried on, and is swamped in
> the competition with the large capitalist, partly because their
> specialized skill is rendered worthless by new methods of
> production.[21]

The 'informal' sector is not, however, solely constituted by those
above who would prefer to work elsewhere. While this is the
impression that most of the literature provides, several pieces of
evidence qualify this.

The first point is the belief that promotion, higher income,
personal independence and prestige are likely to come quicker and to
a greater extent under self-employment. The lowly employee of a

large establishment may (more or less accurately) assess his chances of promotion as slight. Although the firm may have a structured internal labour market requiring certain recruitment and hiring procedures for the benefit of the existing work force, he may be unable to overcome the imposed requirement that higher-level posts are filled from outside — either by expatriates, or through quasi-political manoeuvres. The ambitious employee may not be encouraged at his or her prospects.

Next, regarding income comparisons, it should be noted that by no means all employees in 'formal' enterprises receive more than 'informal' urban or even rural workers. Although one cannot easily assess real standards of living (still less real income), there is evidence that formal jobs begin with fairly low salaries in some countries. Table 6.1 reproduces some data on this. This table confirms that the 'aristocracy of labour' alleged to exist in the 'modern' or 'formal' sector of UDCs' urban labour markets is often a chimera. Skilled craftsmen in many cases earn substantially more than those in hierarchical labour structures in larger enterprises (see chapter 3). In fact a surprisingly large proportion of the urban poor in some cities are government employees. Some 11 per cent of the poor in Malaysia, 12 per cent in Belo Horizonte and 10 per cent in Lima are thus employed.[22] (Definitions of the 'urban poor', of course, differ between cities and over time.) Some cities also have poor inhabitants whose prime source of earnings is agriculture. The problem, referred to earlier, of defining a city is apparent here: if areas suitable for agriculture are included in the definition of a city, different types of poverty will be found. In Bangkok it has been estimated that 29 per cent of the poor have agriculture as their main source of subsistence; in Malaysia and Peru the proportions are 15 per cent and 10 per cent respectively.

There is, furthermore, as was pointed out in chapter 5, greater prestige attached to self-employment than to being an employee in some UDCs. In Ghana and Nigeria, for instance, 'the desire to be your own boss is very strong . . . especially among manual workers'.[23] To achieve this, some will elect to forego self-esteem or social acceptability in the short term, to accumulate money for self-employment later. Another reason for preferring a short spell in such an occupation is that it can serve as a conduit to the 'formal' sector, in so far as the latter enterprises prefer trained over untrained workers. The extent to which this is true will depend upon the strength and spread of internal labour markets. A company which is experiencing a shortage of supervisory or skilled labour will tend to look outside its

Table 6.1: Earnings in Formal and Other Sectors

Place	Date	Currency	Time Reference Day Week Month Year	'Formal' sector Skilled w_s	Unskilled w_u	'Traditional' skilled w_x
Ouagadougou	1969?	CFAfr	w			2,500[a]
Khartoum	1956?	£S	y	300	77[a]	375
Poona	1954	Rs	y	1,200	790	730
Nairobi	mid-1960s	sh	w	40[a]	34[a]	50[a]
Kenya	early 1970s	sh	m	450[a]	225[a]	120[a]
Iseyin	1965-6	£N	m			15[a]
Ibadan	c.1964	£N	m		6.37[a]	11.25[a]
Ibadan	1959-62	£N	m		6.25[a]	8.5[a]
West Bengal	1972-4	Rs	m	250[a]	174[a]	150[a]
Kanpur	1954-6	Rs	m	80[b]	47[b]	48[b]
Bombay	1954	Rs	m	110[a]	71[b]	104[b]
North Central State, Nigeria	1972	£N	y	250[a]	120[a]	157
Kwara State, Nigeria	1972	£N	y			144
Mysore State	1963 or 1964	Rs	d	4.51	4.43	2[a]
Amravati City, India	c.1960	Rs	m			63.46[b]
Sholapur City, India	1938-9	Rs	y	226[b]	157[b]	110[b]
Saugor City, India	1956?	Rs	m			37.5[a,b]
Bangkok	1975	Bt	d	55	29	
Addis Ababa	1968-9	$E	d	2.72[a]	1.20[a]	2.075[a]
W. Bengal	1966	Rs	y	1,383	1,320	747
Ahmedabad	c.1961	Rs	m	159[a]	122[a]	150
Hyderabad	c.1961	Rs	d			2.25
Karimnagar	c.1961	Rs	d			2
Kondapally (village?)	c.1961	Rs	m			100
Bihar State	1968-70	Rs	y			820[b]
Tanzania (all urban)	1971	sh	m	397	151[a]	200

[a] averaged
[b] estimated

Source: J.G. Scoville, 'The Role and Functioning of the Traditional Industrial Sector: Some Preliminary Evidence' (paper presented to International Institute for Labor Studies, 4th World Conference, Geneva, September 1976), pp. 14-15.

own workforce for extra labour. There is very little empirical evidence as yet on whether much hiring of skilled labour from outside the firm does take place. One might expect the contrary tendency, that a company recruits only young school-leavers and trains internally, to work against such hiring. A theoretical view of internal labour markets by Dore is sceptical about the likelihood of large companies hiring experienced workers from outside; he believes that failure to get into such a company when young implies perpetual exclusion.[24] The whole subject of hiring criteria by 'formal' employers and labour mobility between these and other types of enterprise is just beginning to be studied.

Improvements in the 'Informal' Sector

This section will review some recent reinterpretations of urban labour markets before a summary is given of the criticisms presented above. The concept of the 'intermediate' sector was introduced by William Steel in 1976.[25] This was the name given to a set of enterprises which were larger than what were typically thought to be 'informal' enterprises, yet were not big enough to be considered 'formal'. Based on rare empirical material from Ghana, Steel's analysis attempts to recognise the contribution, in terms of increasing employment and output, made by such people as carpenters, seamstresses and hairdressers. The criteria he uses to differentiate the three sectors are shown in Table 6.2. Using these criteria, Steel estimates that in Ghana the 'intermediate' sector accounts for about 40 per cent of total manufacturing employment, with the 'informal' sector dominating commerce and services.

It is clear at once that the innovation represented by the 'intermediate' sector helps to classify the data which one might collect. Rather than maintaining a crude dichotomy of all enterprises, this at least allows some middle group to be recognised. But might this threefold classification not as easily confuse as illuminate? Essentially the 'intermediate' sector is another exercise in *ex post facto* comparative statics.[26] Whether or not the characteristics ascribed to different 'sectors' were necessary and sufficient characteristics for them to reside in one 'sector' but not in any other, or whether these characteristics could be found in enterprises in other 'sectors' too, is not clear. Furthermore, whether or not 'intermediate' enterprises set out thus, or began as 'informal' enterprises, is a question which this typology is ill-equipped to answer.

Table 6.2: Criteria to Distinguish Four Urban 'Sectors'

Sector	Wage	Productivity	Employment	Capital	Size, technology, organisation
Modern	High (i.e. minimum wage)	High	Wage labour	Capital-intensive	Large (modern technique
Intermediate	Low	Medium	Wage, also apprentice, family, self-employment	Some fixed capital, but relatively labour-intensive	Small (may have some employees)
Informal	Low	Low	Self-employment Non-wage family	No fixed capital	Very small (no formal business organisation
Unemployed	No earned income	—	Seeking a job	—	—

Source: Steel, p. 25.

Comparative Statics and Dynamics

The arguments put forward in chapter 5 with respect to the 'informal' sector, and earlier in chapter 6 with respect to the 'intermediate' sector must now be drawn together. The important questions are whether classification — ascription of enterprises to categories — is in any sense a substitute for dynamic analysis; and whether dynamic analysis is facilitated by statics. The conclusions one must draw at the present state of knowledge are that dynamic analysis of urban labour markets yields more light on how urban labour markets hang together, but that, in the absence of such knowledge, classificatory work is useful for helping to pose the questions used in that later dynamic work. Classification *per se* is not a sufficient basis for development research. But such work may provide the basis for analysing movements of labour and capital between categories, as Steel's studies in Ghana suggest:

> Unsuccessful job-seekers and laid-off workers must go into business to earn enough income to survive . . . Those who meet with some success may use family members for assistance, and

eventually engage part-time help or apprentices . . . Apprentices provide the dynamic mechanism for perpetuating this [the intermediate] model, since they can use their acquired skill to go into business for themselves . . . Some small-scale firms flourish sufficiently to increase the number of apprentices, eventually having to hire additional masters to help supervise the apprentices . . . part-time workers taken on to meet particular orders may become full-time wage workers as the stream of orders becomes steady. Thus the firm enters the modern sector, and as the employer gains business experience, it may grow into a large-scale establishment.[27]

As the equivocation of this account shows, however, Steel agrees that in-depth studies are needed before the actual evolutionary process and the responses to different policies and inter-firm competition can be adequately understood. This transitional model of Steel's is Rostovian in the sense that it outlines stages that do not contain inevitable forces leading on to the next stage; and it could certainly benefit from evidence on the dynamics of movement (or non-movement) between stages. A similar judgement is that of Scoville, whose work on urban labour markets' interaction began with the analysis of entry into skill-hierarchies in Afghanistan.[28] He argues that, 'even as the elaboration of additional sectors gets us away from the spurious concreteness of dualistic models, we must bear in mind that sectors are only hazily defined in practice'.[29]

The conclusions which one can draw, given the present state of knowledge of urban labour-market interactions, about small-scale enterprises are as follows. The evolution of the concept of the 'informal' sector as a heuristic device in development studies was a useful antidote to the overly rigid application of unemployment-based studies. Rather than spending time and effort in measuring varieties of unemployment, it had become apparent that recognising the different forms of labour-use in cities was a more promising path to follow. Once this was done, research was directed to empirical estimation of the number of people engaged in, and the earnings derived from, such 'informal' units. But this work was compromised, it was argued, by the unsatisfactory theoretical basis of the 'informal' sector itself. Many researchers had a different idea of what constituted this sector, and, in the absence of a commonly accepted set of criteria, it was inevitable that conflicting research results and policy proposals would ensue. Yet even the provision of common definitions for researchers would be insufficient to illuminate the study of these people's work and earnings conditions, critics argued. For inherent in adherence to the 'informal' approach were a number of analytical shortcomings. The three most important of these were: first, the assumption of sufficient homogeneity of units in the sector to permit undifferentiated analysis to take place; second, the absence of clues regarding the ways these units interacted with others, similar or different, in the urban economic milieu; and third, the absence of a way of changing from a comparative statics framework to a dynamic framework. What is entailed in this last point is the need for supplementary pieces of analysis to explain how enterprises with characteristics usually ascribed to 'informal' units become larger or smaller. Of particular importance here is the need to explain the tendency for many small operations to proliferate under the ownership of one person. The few studies available recording distribution of small enterprises by number employed (assuming criteria are established to determine what precisely constitutes 'employment') show a marked tendency for the one-man scale to be retained. Further empirical investigation, combined with experimental theoretical bases upon which to arrange the data, is the urgent priority.

Notes

1.   O. Olakanpo, 'Distributive Trade: A Critique of Government Policy', *Nigerian Journal of Economic and Social Studies*, Vol. 5, No. 2, July 1963, pp. 237-46.
2.   M. Kolawole, 'Food Retailing in Nigeria', *ODU*, No. 10, July 1974, pp. 108-17.
3.   S.U. Sethuraman, 'Urbanization and Employment: A Case Study of Djakarta', *International Labour Review*, Vol. 112, Nos. 2-3, August-September 1975, pp. 191-206.
4.   L.G. Reynolds, 'Economic Development with Surplus Labour: Some Complications', *Oxford Economic Papers*, Vol. 21, No. 1, March 1969, p. 91.
5.   T.W. Merrick, 'Employment and Earnings in the Informal Sector in Brazil: The Case of Belo Horizonte', *Journal of Developing Areas*, Vol. 10, No. 3, April 1976, pp. 337-54. Quotation is from p. 351.
6.   D.C. Rao, 'Urban Poverty Groups', in H. Chenery *et al.* (eds.), *Redistribution with Growth* (London: Oxford University Press for IBRD, 1974), pp. 136-57.
7.   S.W. Sinclair, 'Ease of Entry into Small Scale Trading in African Cities: Some Case Studies from Lagos', *Manpower and Unemployment Research*, Vol. 10, No. 1, April 1977, pp. 79-90.
8.   T. Adekokunna, 'The Market for Foodstuffs in Western Nigeria', *ODU*,

Vol. 3, No. 5, April 1970, pp. 71-86. Quotation is from p. 80.
9.   K. Little, *African Women in Towns* (London: Cambridge University Press, 1973). pp. 50-2.
10. K. Little, *West African Urbanization* (London: Cambridge University Press, 1965), pp. 48-57.
11. D.C. Lewis, 'The Limitations of Group Action Among Entrepreneurs: The Market Women of Abidjan, Ivory Coast', in N.J. Hafkin and E.G. Bay (eds.), *Women in Africa: Studies in Social and Economic Change* (Stanford: Stanford University Press, 1976), pp. 135-56.
12. L.R. Vagale, 'Anatomy of Traditional Markets in Nigeria: Focus on Ibadan City' (The Polytechnic, Ibadan: Town Planning Dept., 1974), pp. 8-10.
13. G. Papanek, 'The Poor of Jakarta', *Economic Development and Cultural Change*, Vol. 24, No. 1, October 1975, p. 10.
14. R.J. Bromley and R. Symanski, 'Marketplace Trade in Latin America', *Latin American Research Review*, Vol. 9, No. 3, Fall 1974, p. 11.
15. J. Breman, 'A Dualistic Labour System? A Critique of the "Informal Sector" Concept', *Economic and Political Weekly* (Bombay), 27 November, 4 December and 11 December 1976 (in three parts).
16. P. Streefland, 'The Absorptive Capacity of the Urban Tertiary Sector in Third World Countries', *Development and Change*, Vol. 8, No. 3, July 1977, pp. 293-305.
17. K. Newcombe, 'From Hawkers to Supermarkets: Patterns of Food Distribution', *Ekistics*, Vol. 43, No. 259, June 1977, pp. 336-41.
18. P. Peek and P. Antolinez, 'Migration and the Urban Labour Market: The Case of San Salvador', *World Development*, Vol. 5, No. 4, 1977, pp. 291-302.
19. K. Shaefer, *Sao Paulo: Urban Development and Employment* (Geneva: ILO, 1976), pp. 46-59.
20. J. Van Velsen, 'Urban Squatters: Problem or Solution?', in D. Parkin (ed.), *Town and Country in Central and Eastern Africa* (Oxford: Oxford University Press for the International African Institute, 1975), pp. 294-307.
21. K. Marx, 'The Manifesto of the Communist Party'.
22. IBRD, 'On the Statistical Mapping of Urban Poverty and Employment' (Washington DC: World Bank Staff Working Paper, No. 227, January 1976), p. 27.
23. P. Kennedy, 'Cultural Factors Affecting Entrepreneurship and Development in the Informal Economy in Ghana', *Institute of Development Studies Bulletin*, Vol. 8, No. 2, February 1976, pp. 17-21.
24. R.P. Dore, 'The Labour Market and Patterns of Employment in the Wage Sector of LDCs: Implications for the Volume of Employment Created', *World Development*, Vol. 2, Nos. 4 and 5, April-May 1974, pp. 1-7.
25. W.F. Steel, 'Empirical Measurement of the Relative Size and Productivity of Intermediate Sector Employment: Some Estimates from Ghana', *Manpower and Unemployment Research*, Vol. 9, No. 1, April 1976, pp. 23-31. Greater detail is provided in W.F. Steel, *Small-Scale Employment and Production in Developing Countries: Evidence from Ghana* (New York: Praeger, 1977), Ch. 7.
26. S.W. Sinclair, 'The "Intermediate" Sector in the Economy', *Manpower and Unemployment Research*, Vol. 9, No. 2, November 1976, pp. 55-60.
27. W.F. Steel, 'Static and Dynamic Analysis of the Intermediate Sector: A Synthesis', *Manpower and Unemployment Research*, Vol. 10, No. 1, April 1977, pp. 73-8. Quotation is from p. 76.
28. J.G. Scoville, 'Afghan Labor Markets: A Model of Interdependence', *Industrial Relations*, Vol. 13, No. 3, October 1974, pp. 274-87.

29.   J.G. Scoville, 'The Role and Functioning of the Traditional Industrial Sector: Some Preliminary Evidence' (paper presented to International Institute for Labor Studies, 4th World Conference, Geneva, September 1976), p. 8.

7 CONCLUSIONS

This book has been concerned with labour markets within the fast-growing cities of the third world. The reasons for migrants leaving farms and small towns, and arriving in a major city, possibly through a sequence of moves to increasingly large towns, were examined in chapter 2. Their reception, the subject of chapter 3, consists of various urban support networks and processes of looking for work. The reasons for many migrants not being employed in 'formal' firms were reviewed in chapter 4. Chapter 5 traced the development of the 'informal' sector concept and found it wanting, and chapter 6 then considered a few alternatives and put forward imperatives for better analysis of the urban working poor. Questioning the proposals of academics wedded to a typological approach to urban labour markets and shanty towns ('we need an improved taxonomy of urban situations . . .'[1]) the chapter tries to show the fluidity of labour mobility. This is what links labour markets hitherto too rigidly set apart from one another in much of the literature.

Although some proportion of the growth of cities is temporary, and thus in a sense self-curing, steps can be taken to mitigate the consequences of urban hypertrophy in the shortrun. The reasons for believing that some degree of cities' growth is temporary were adduced in chapters 1 and 2: that many migrants, particularly in South and East Asia, and East and West Africa, plan to retire elsewhere — typically in their town or village of origin — and that many will, in the next 20 years or so, leave the centre of the city as it is presently structured in favour of residence in newly-established satellite towns, or in outlying areas of existing towns.

The chief constraints upon short-term improvements in cities are administrative bottlenecks and low tax revenues. It is generally agreed that the capacity for sound municipal government is lacking: 'Without changes in their administrative structure, few cities will be able to handle adequately the problems of the coming decades — even if sufficient resources are available.'[2] Policies which discriminate against the poor (setting rents and fares too high, requiring frequent and large movements in the residential areas of the poor, setting public housing standards too high) still reflect the over-formalised attitudes prevalent before Turner and Mangin proposed that shanty settlements

Table 7.1: Urban Per Capita Public Expenditure and Revenues ($ US)

City	Year	Per capita expenditure	Per capita revenues
Africa			
Blantyre	1965-6	8.12	8.91
Dar-es-Salaam	1965	11.74	11.55
Fort Lamy	1970	14.56	11.10
Freetown	1963-4	15.35	15.35
Lusaka	1967	—	31.00
Nairobi	1965	42.99	42.96
Far East			
Bangkok	1968	10.16	10.56
Djakarta	1970-1	4.70	5.74
Manila	1964	—	16.54
Seoul	1970	29.97	31.22
Singapore	1969	115.05	143.14
South Asia			
Bombay	1970	13.65	11.83
Calcutta	1967-8	4.32	4.04
Madras	1967-8	6.49	5.90
Colombo	1968	10.89	12.26
Latin America			
Belo Horizonte	1968	11.57	14.52
Rio de Janeiro	1968	82.32	79.06
Sao Paulo	1968	38.08	38.09
Bogota	1969	28.16	28.16
Cali	1969	13.28	11.44
Mexico	1963	23.51	27.07
Panama City	1968	9.51	8.42
Rich countries			
New York City	1969	888.50	828.80
Stockholm	1968	1,391.06	1,357.66

Source: R.S. Smith, 'Financing Cities in Developing Countries', *IMF Staff Papers*, Vol. 21, No. 2, July 1974, pp. 329-88.

could be seen as 'solutions' rather than 'problems'.

Tax revenue per capita for cities is not only low, but naturally cannot increase substantially unless the wherewithal of urban residents to pay increases first. Table 7.1 shows the small fraction of per capita revenues received in UDCs contrasted with two cities in DCs. Even when price differences are taken into account, it can be seen that there is a considerable disparity in what cities can afford to provide as collective public goods and services. One public provision of particular importance for the poor is housing. UDCs with relatively high income per capita have high proportions of the housing stock provided privately — 45 per cent of Hong Kong and Singapore housing is privately owned — but poorer countries are more dependent upon public provision.[3] Policies that provide low-cost housing public would also create urban income. Multipliers from construction in UDCs average 2, so that in Colombia some 7 jobs are created for each $10,000 spent on dwelling construction; in Korea 14 jobs.[4]

None of these important subjects can be dealt with satisfactorily here. The one aspect of cities chosen for scrutiny — the pattern of labour markets — has itself only been touched upon. Much more work is needed to illuminate the work-patterns and earnings of the various components of the urban poor.

Notes

1. J. Vincent, 'Urbanization in Africa', *Journal of Commonwealth and Comparative Politics*, Vol. 14, No. 3, November 1976, pp. 286-98.
2. 'The Task Ahead for the Cities of the Developing Countries' (Washington DC: World Bank Staff Working Paper No. 209, July 1975), p. 57.
3. O.F. Grimes, Jr., *Housing for Low Income Urban Families* (London: Johns Hopkins Press for IBRD, 1976), p. 13.
4. W. Paul Strassmann, 'Measuring the Employment Effects of Housing Policies in Developing Countries', *Economic Development and Cultural Change*, Vol. 24, No. 3, April 1976, pp. 623-32.

INDEX

Printed in the United States
by Baker & Taylor Publisher Services